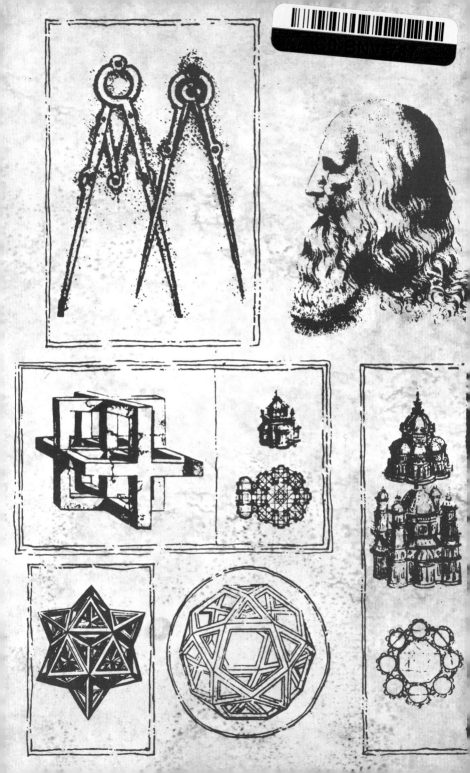

DICTIONARY OF PHYSICS AND MATHEMATICS ABBREVIATIONS
SIGNS AND SYMBOLS

DICTIONARY OF PHYSICS AND MATHEMATICS ABBREVIATIONS
SIGNS AND SYMBOLS

SIMON AND SCHUSTER TECH LIBRARY

RAJ MEHRA, EDITOR

SIMON AND SCHUSTER • NEW YORK

A Simon and Schuster Tech Library Book
Published by Simon and Schuster
Technical and Reference Book Division
1 West 39th Street
New York, N.Y. 10018

Library of Congress Catalog Card Number 65-18843

DPMA

Copyright © 1965 by Western Publishing Company, Inc.
Reprinted by arrangement with Western Publishing Company, Inc.
First Simon and Schuster Tech Library printing 1971.

All rights reserved including the right of reproduction
in whole or in part in any form.

Printed in the U.S.A.

PREFACE

LANGUAGE is an instrument of communication, and as such, develops and becomes more symbolic as man's thoughts grow increasingly sophisticated and complex. Scientists and engineers, under the compelling necessity to communicate with each other in an expedient yet precise way, have used their own inventive spirit in a quest for linguistic means to express their complex thoughts. Traditionally, they have built their own words and symbols out of bits furnished by existing languages. The purest and probably the oldest of all sciences, mathematics, solved the problem by resorting to letters, symbols and equations, each representing a precisely defined concept. The meaning of all these symbols, their rules and interrelations, and their proper uses must become an integral part of every student of the physical sciences.

There has been an explosive and diversified growth in the sciences during this century, accompanied by further specialization within each discipline, and an increasingly intricate interaction between all disciplines. Even within a given field, such as physics or mathematics, the extensive use of abbreviations and an abbreviated terminology can lead to misinterpretation within different specialties. These difficulties can be greatly magnified for the physicist attempting to communicate with the engineer, the scientist with the nonscientist, etc.

Abbreviations, signs and symbols representing concepts, quantities and operations, are intrinsic to the language of science, and as with any language, its effectiveness depends upon concensus as to form and meaning. In general, standards for abbreviations either do not exist, or are not widely accepted, which means that several different abbreviations may be used for a single phrase or word. Standardization is never easy; nevertheless, standards are established, and while they do not resolve all existing disagreements, they do tend to minimize the number of conflicts and provide a stable base for further refinement and development.

Although this dictionary does not attempt to establish standards, it is hoped the cataloging of abbreviations, signs and symbols in physics and mathematics will help bring about a uniformity of notation.

PREFACE

The organization of this volume has been planned for ease of use. Each of the three major categories—abbreviations, signs and symbols—is grouped into a separate section. With the exception of mathematical signs and symbols, each section is divided into two parts: the first part is an alphabetical listing of the abbreviation or letter symbol, and the second part is an alphabetical listing of the term itself. One may find the abbreviation or letter symbol by looking for the term, or the term by referring to the abbreviation or letter symbol.

The distinction between abbreviations and letter symbols is as follows:

(1) An abbreviation is a shortened form of a word or term generally appearing in English letters in roman type, but occasionally using numerals or Greek letters.

(2) Letter symbols are generally single letters, from any of a number of alphabets, representing numerical and physical quantities, often appended with subscripts or superscripts to qualify their meaning.

Where more than one abbreviation or symbol is in common use for a given term, both or all are included. The sources are given for all entries with the exception of those in accepted American usage.

The international and foreign standards groups whose recommendations are included in this volume are the International Union of Pure and Applied Physics, the International Organization for Standardization, the International Electrotechnical Commission, the International Union of Pure and Applied Chemistry, the British Standards Institution, the Canadian Standards Association, and the British Institution of Electrical Engineers. Entries from such sources are always indicated.

I wish to express my deep gratitude for the untiring efforts of Dolly Nielsen in the revision for the reprint of this dictionary.

Raj Mehra
New York

CONTENTS

Preface v

1. ABBREVIATIONS FOR USE IN TEXT
Introduction 1
Alphabetically by abbreviation 5
Alphabetically by definition 65

2. LETTER SYMBOLS
Introduction 125
Alphabetically by symbol 127
Alphabetically by definition 194

3. MATHEMATICAL SIGNS AND SYMBOLS 294

4. ABBREVIATIONS FOR SCIENTIFIC AND LEARNED SOCIETIES
Introduction 305
Alphabetically by abbreviation 306
Alphabetically by name 309

5. ABBREVIATIONS FOR GOVERNMENT AND MILITARY DEPARTMENTS, AGENCIES, AND OFFICES
Introduction 312
Alphabetically by abbreviation 313
Alphabetically by name 317

6. TABLE OF CHEMICAL ELEMENTS 321

7. PERIODIC TABLE OF THE ELEMENTS 324

8. SOURCES OF STANDARDS AND DOCUMENTS 326

ABBREVIATIONS FOR USE IN TEXT

ABBREVIATIONS are as much a part of the scientific language as the terms they represent. Their use permits greater economy in communication, and they have consequently become increasingly valuable in saving both time and space in the preparation of scientific papers.

The technical abbreviations compiled in this section are those commonly used by the international scientific community. Abbreviations for less technical terms are from the best available sources; abbreviations for days and months, for example, are those given in the United States Government Printing Office "Style Manual."

The following rules govern the use of abbreviations:

(1) The only punctuation used with abbreviations is a final period, and in most cases, this is used only when the abbreviation of one word implies another. However, even where punctuation is concerned there is no general agreement. For instance, the abbreviation "in." (with a period) for "inch" is used in the United States, whereas the British use "in" (without the period).

There are also many cases where the final period has become unnecessary, such as in the abbreviations for "log", "sin", "tan", etc.

(2) The same abbreviation is used for both the plural and singular forms of a word. Though exceptions do exist, different syntactical uses or forms of the same word generally have the same abbreviation. Parenthetical notes are appended to the definitions in this section, interpreting, clarifying or limiting the use of that abbreviation. Two notes require further explanation:

(adjective) .... indicates that the abbreviation listed is applicable only to the adjectival form of the word and implies the existence of a different abbreviation for its noun form.

(adjective only) .... indicates that only the adjectival form of the word is abbreviated, never the noun.

Often technical publications erroneously change the abbreviation for the different forms of a word where it should remain the same for all cases. These are listed independently.

(3) When a group of letters is used to abbreviate a phrase consisting of more than one word, the separate letters are not to be re-

garded as necessarily valid abbreviations for the individual words. However, abbreviations for single words may be combined when a group of words is to be abbreviated.

Some sources deprecate the use of abbreviations for phrases, preferring the combination of the abbreviations for the separate words. Current use favors the shorter abbreviation for the full term if such an abbreviation is clear and free from ambiguity.

(4) Subscripts are not usually appended to abbreviations, but there are exceptions. Some sources, among them the American Standards Association, also discourage the textual use of exponents in place of abbreviations for "square" and "cube" to avoid confusion with footnote reference numbers. The American Standards Association notes, however, that these exponents are expedient in tables and "are sometimes difficult to avoid in text." Traditional practice, on the other hand, prefers the use of exponents to word abbreviations.

(5) Though the standard designations for chemical elements are not formal abbreviations, they are included in this section because they are used as such in textual material.

(6) Generally, short words are not abbreviated; therefore only those established by tradition have been included. When units of measurements are too short to be abbreviated, they are listed alphabetically with the instruction "(spell)."

(7) An abbreviation should be used only when its meaning is clear; otherwise it should be spelled to avoid ambiguity. In most cases, only one abbreviation for any particular term is in standard American usage. Where more than one abbreviation has achieved general acceptance, all are listed, with no indication for preference. The source(s) of such entries are not given. Thus, an absence of reference following the definition indicates general American acceptance of the abbreviation.

Specific preferences of different societies, associations and international bodies are given in brackets, following the definition, even where the form is exactly the same as standard American usage.

The sources used in this section are noted as follows:

[AAAS]	American Association for the Advancement of Science
[AAS]	American Astronomical Society
[ACS]	American Chemical Society
[AEC]	Atomic Energy Commission
[AGU]	American Geophysical Union
[AIP]	American Institute of Physics
[ASA]	Americans Standards Association

DPMA

[British]	British Standards Institution
[British, I.E.E.]	British Institution of Electrical Engineers[1]
[Canadian]	Canadian Standards Association
[IEC]	International Electrotechnical Commission
[International]	International acceptance[2]
[ISO]	International Organization for Standardization
[IUPAC]	International Union of Pure and Applied Chemistry
[IUPAP]	International Union of Pure and Applied Physics
[LRL]	Lawrence Radiation Laboratory of the Atomic Energy Commission
[NBS]	National Bureau of Standards

[1] Identical recommendations from this source and the British Standards Institution are credited only to the British Standards Institution, as its recommendation implies general British use.

[2] Indicates only those decimal multiples and submultiples of metric system units that have gained international acceptance.

ABBREVIATIONS FOR USE IN TEXT
Alphabetically by Abbreviation

1/W	one-way	
2-D	two-dimensional	
2/W	two-way	
3-D	three-dimensional	
4-D	four-dimensional	
a	ampere (combining form) [AAAS]	
a	atto ($=10^{-18}$) (prefix) [International]	
a	year [ISO]	
a	year [IUPAP]	
A	ampere	
A	ampere [British]	
A	ampere [ISO]	
A	ampere [IUPAC]	
A	ampere [IUPAP]	
A	angstrom [ASA]	
A	angstrom [Canadian]	
Å	angstrom	
Å	angstrom [British]	
Å	angstrom [ISO]	
Å	angstrom [IUPAC]	
Å	angstrom [IUPAP]	
A_s	atmosphere, standard [AIP]	
abamp	absolute ampere	
ABRES	Advanced Ballistic Re-entry Studies	
abs	absolute	
abs	absolute [Canadian]	
abs	absolute value of	
abs.	absolute [British]	
absc	abscissa	
absorp	absorption	
abstr	abstract	
a.c.	alternating current [British]	
a-c	alternating-current (adjective)	
a-c	alternating-current (adjective only) [Canadian]	
ac	acute	
ac	alternating current	
Ac	actinium	
accel	acceleration	

ACM	atom connectivity matrix	
ACMCP	atom connectivity matrix characteristic polynomial	
acous	acoustics	
A.D.	anno Domini (in the year of our Lord)	
AD	average deviation	
adb	acceleration decibel	
add.	addition	
ad fin.	ad finem (at the end)	
ad inf.	ad infinitum (to infinity)	
ad init.	ad initium (at the beginning)	
ad int.	ad interim (meanwhile)	
adj	adjacent	
adj.	adjective	
ad lib.	ad libitum (at pleasure)	
ad loc.	ad locum (at the place)	
ADP	automatic data processing	
adv.	adverb	
AE	activation energy	
aerodyn	aerodynamic	
af	audio frequency	
af	audio-frequency (adjective) [AIP]	
AF	attenuation factor	
Ag	silver	
aggr	aggregate	
agt	agent	
A-h	ampere-hour [AIP]	
Ah	ampere-hour	
Al	aluminum	
alg	algebra	
algo	algorithm	
alk	alkali	
alky	alkalinity	
alt	altitude	
aly	alloy	
a.m.	ante meridiem (before noon)	
am	amplitude, an elliptic function [Canadian]	
am.	amplitude, an elliptic function	
Am	americium	
Am.	American	
AM	amplitude modulation	
AMA	actual mechanical advantage	
Amer Std	American Standard	

amor	amorphous	
amp	ampere [British]	
amp	ampere [Canadian]	
amp-hr	ampere-hour [Canadian]	
AMS	Applied Mathematics Series [NBS]	
amt	amount	
amu	atomic mass unit	
anal.	analysis	
ant.	antonym	
antilog	antilogarithm	
antilog	antilogarithm [British]	
antilog	antilogarithm [Canadian]	
antitrig	antitrigonometric	
AP	arithmetic progression	
app	appendix	
appl	application	
approx	approximate	
approx.	approximate [British]	
approxn	approximation	
Apr.	April	
aq. bull.	aqua bulliens (boiling water)	
aq. dest.	aqua destillata (distilled water)	
Ar	argon	
arc cos	inverse cosine	
arc cos	inverse cosine [British]	
arccos	inverse cosine [ISO]	
arccos	inverse cosine [IUPAP]	
Arc cos	principal value of inverse cosine	
arc cosec	inverse cosecant [British]	
arccosec	inverse cosecant [ISO]	
arccosec	inverse cosecant [IUPAP]	
arc cosech	inverse hyperbolic cosecant [British]	
arc cosh	inverse hyperbolic cosine [British]	
arc cot	inverse cotangent	
arc cot	inverse cotangent [British]	
arccot	inverse cotangent [ISO]	
arccot	inverse cotangent [IUPAP]	
Arc cot	principal value of inverse cotangent	
arc coth	inverse hyperbolic cotangent [British]	
arc csc	inverse cosecant	
Arc csc	principal value of inverse cosecant	
arcctg	inverse cotangent [ISO]	

arcctg	inverse cotangent [IUPAP]	
arc ctn	inverse cotangent	
Arc ctn	principal value of inverse cotangent	
arcosech	inverse hyperbolic cosecant [ISO]	
arcosech	inverse hyperbolic cosecant [IUPAP]	
arcosh	inverse hyperbolic cosine [ISO]	
arcosh	inverse hyperbolic cosine [IUPAP]	
arcoth	inverse hyperbolic cotangent [ISO]	
arcoth	inverse hyperbolic cotangent [IUPAP]	
arc sec	inverse secant	
arc sec	inverse secant [British]	
arcsec	inverse secant [ISO]	
arcsec	inverse secant [IUPAP]	
Arc sec	principal value of inverse secant	
arc sech	inverse hyperbolic secant [British]	
arc sin	inverse sine	
arc sin	inverse sine [British]	
arcsin	inverse sine [ISO]	
arcsin	inverse sine [IUPAP]	
Arc sin	principal value of inverse sine	
arc sinh	inverse hyperbolic sine [British]	
arc tan	inverse tangent	
arc tan	inverse tangent [British]	
arctan	inverse tangent [ISO]	
arctan	inverse tangent [IUPAP]	
Arc tan	principal value of inverse tangent	
arc tanh	inverse hyperbolic tangent [British]	
arctg	inverse tangent [ISO]	
arctg	inverse tangent [IUPAP]	
arctgh	inverse hyperbolic cotangent [IUPAP]	
ARD	automatic recorder of static deformation	
arg	argument	
arg	argument of	
arg	argument of [ISO]	
arg	argument of complex number [British]	
arith	arithmetic	
arsech	inverse hyperbolic secant [ISO]	
arsech	inverse hyperbolic secant [IUPAP]	
arsinh	inverse hyperbolic sine [ISO]	
arsinh	inverse hyperbolic sine [IUPAP]	
art.	article	
artanh	inverse hyperbolic tangent [ISO]	

artanh	inverse hyperbolic tangent [IUPAP]
artgh	inverse hyperbolic tangent [IUPAP]
As	arsenic
a.s.a.	angle, side, angle
asb	apostilb
asst	assistant
astron	astronomical
astron	astronomy
astrophys	astrophysics
asym	asymmetric
asymp	asymptote
at.	atomic
At	ampere-turn
At	astatine
AT	ampere-turn [British]
AT	atomic time
atm	atmosphere
atm	atmosphere [British]
atm	atmosphere [Canadian]
atm	atmosphere [IUPAP]
atm	atmosphere, standard [British]
atm	atmospheric
atm.	atmospheric [British]
At/m	ampere-turns per meter
atm press.	atmospheric pressure
at. no.	atomic number
A-turn	ampere-turn [LRL]
at. vol	atomic volume
At/Wb	ampere-turns per weber
at. wt	atomic weight
at. wt	atomic weight [Canadian]
at. wt.	atomic weight [British]
a.u.	astronomical unit [AAS]
au	atomic unit
Au	gold (aurum)
AU	astronomical unit
AU	astronomic unit [ISO]
Aug.	August
auto	automatic
av	average [AIP]
avdp	avoirdupois
avdp	avoirdupois [Canadian]

avg	average	
avg	average [Canadian]	
awu	atomic weight unit	
ax.	axis	
Ax	axiom	
az	azimuth	
az	azimuth [Canadian]	
b	bar [British]	
b	bar [IUPAC]	
b	barn	
b	barn [IUPAP]	
b	bel (combining form)	
b.	boils at	
B	boron	
°B	degree Baumé [AIP]	
Ba	barium	
bal	balance	
bal	ballistic	
bat.	battery	
bb	black body	
b.c.	body-centered (crystals)	
B.C.	before Christ	
b.c.c.	body-centered cubic (crystals)	
bcc	body-centered cubic (crystals) [AIP]	
BCD	binary coded decimal	
BCS theory	Bardeen-Cooper-Schrieffer theory of superconductivity	
bdy	boundary	
Be	beryllium	
Belg.	Belgian	
bev	billion electron volts	
BeV	billion electron volts [AIP]	
BF	brittle fracture	
Bhn	Brinell hardness number	
Bhn	Brinell hardness number [Canadian]	
b.h.p.	brake horsepower [British]	
bhp	brake horsepower [Canadian]	
Bi	biot	
Bi	bismuth	
biax	biaxial	
bicond	biconditional	
bin.	binary	

binom	binomial
binom exp	binomial expansion
binom th	binomial theorem
biochem	biochemistry
bipyr	bipyramidal
birect	birectangular
bis	bisector
bk	book
Bk	berkelium
blws	bellows
BONUS	boiling nuclear superheat (reactor)
b.p.	boiling point [British]
bp	boiling point
bp	boiling point [Canadian]
B.P.	before present
BPR	beryllium physics reactor
Br	bromine
Brit.	British
Br Std	British Standard
B.S.	British Standard [British]
Btu	British thermal unit
Btu	British thermal unit [British]
Btu	British thermal unit [Canadian]
BUE	built-up edge
bull.	bulletin
BV	billion volts
BV	breakdown voltage
BW	bandwidth
BWR	boiling-water reactor
c	candle
c	candle [Canadian]
c	centi ($=10^{-2}$) (prefix) [International]
c	curie
c	curie [British]
c	cycle (combining form)
c	cycles per second [ASA]
c	cycles per second [Canadian]
c	cycles per second (combining form)
c.	circa (about)
C	capacitor
C	carbon
C	Celsius

C	....	centigrade
C	....	coulomb
C	....	coulomb [British]
C	....	coulomb [IUPAC]
C	....	coulomb [IUPAP]
C	....	degree centigrade [ASA]
C	....	degree centigrade [Canadian]
C	....	hundred
C	....	hundred [Canadian]
°C	....	degree Celsius
°C	....	degree Celsius [IUPAC]
°C	....	degree Celsius (temperature value) [British]
°C	....	degree centigrade
C°	....	Celsius degree (temperature difference) [AAAS]
ca.	....	circa (about)
Ca	....	calcium
cal	....	calibrate
cal	....	calorie
cal	....	calorie [British]
cal	....	calorie [Canadian]
cal	....	calorie [IUPAC]
cal	....	calorie [IUPAP]
Cal	....	Calorie (large)
calbr	....	calibration
calc	....	calculate
calc	....	calculated
calc	....	calculator
calc	....	calculus
calc.	....	calculated [British]
Can.	....	Canadian
canc	....	cancel
cantil	....	cantilever
cap.	....	capacitor
cap.	....	capacity
cap.	....	capital (letter)
card.	....	cardinal
cat.	....	catalog
cat.	....	catalyst
cat.	....	category
caust	....	caustic
cav	....	cavity
cb	....	center of buoyancy

Cb	columbium
C-bomb	cobalt bomb
c.c.	complex conjugate
cc	carbon copy
cc	cubic centimeter
cc/min/kg	cubic centimeters per minute per kilogram
CCW	counterclockwise
c.d.	current density [British]
cd	candela
cd	candela [British]
cd	candela [ISO]
cd	candela [IUPAC]
cd	candela [IUPAP]
cd	current density
Cd	cadmium
CDD	critical degree of deformation
cd/m^2	candelas per square meter
Ce	cerium
CE	circular error
CEA	circular error average
cel	celestial
celescope	celestial telescope
cemf	counter electromotive force
cemf	counter electromotive force [Canadian]
cent.	centrifugal
cent.	century
CEP	circular error probable
c.f.	centrifugal force
cf	center of flotation
cf.	confer (compare)
Cf	californium
CF	coriolis force
c.g.	center of gravity [British]
cg	center of gravity
cg	centigram
cg	centigram [Canadian]
cgs	centimeter-gram-second
cgs	centimeter-gram-second [Canadian]
cgse	centimeter-gram-second electrostatic
cgsm	centimeter-gram-second electromagnetic
cgsu	centimeter-gram-second unit
CGT	congestion theory

chan	channel	
chap.	chapter	
chd	chord	
chem	chemical	
chem	chemical [Canadian]	
chem	chemist	
chem	chemistry	
chg	charge	
chk	check	
chu	centigrade heat unit	
ch χ	Chapman function	
Ci	cosine integral function	
Ci	curie [AIP]	
C. I.	Color Index	
cir	circle	
cir	circular	
cir	circular [Canadian]	
CIRA	COSPAR International Reference Atmosphere	
circ	circulation	
circ	circumference	
cir mils	circular mils	
cir mils	circular mils [Canadian]	
ckt	circuit	
cl	centiliter	
cl	centiliter [Canadian]	
cl	class	
cl	close	
Cl	chlorine	
CL	complex loading	
clar	clarification	
clar	clarify	
class.	classification	
CLT	central limit theorem	
c.m.	center of mass	
cm	centimeter	
cm	centimeter [British]	
cm	centimeter [Canadian]	
cm	centimeter [IUPAP]	
Cm	curium	
c/m^2	candles per square meter	
cm^2	square centimeter	
cm^2	square centimeter [British]	

DPMA

cm²	square centimeter [Canadian]	
C/m²	coulombs per square meter	
cm³	cubic centimeter	
cm³	cubic centimeter [British]	
cm³	cubic centimeter [Canadian]	
cm Hg	centimeter of mercury	
cm/min	centimeters per minute	
cm oil	centimeter of oil	
cm/sec	centimeters per second	
cn	cosine of the amplitude, an elliptic function	
cn	cosine of the amplitude, an elliptic function [Canadian]	
CNA	cosmic noise absorption	
CNP	celestial north pole	
Co	cobalt	
coax	coaxial	
coef	coefficient	
coef	coefficient [Canadian]	
coeff	coefficient [AIP]	
coeff.	coefficient [British]	
coh	coherent	
col	column	
coll	colloid	
coll	colloidal	
collm	collimator	
colog	cologarithm	
colog	cologarithm [Canadian]	
com	common	
comb.	combination	
comb.	combine	
comb.	combustion	
comm	communication	
comm	commutator	
Comm.	Committee	
comp	companion	
comp	compass	
comp	component	
comp	composite	
comp	composition	
comp	compound	
comp	computation	
compen	compensate	

compil	compilation	
compl	complement	
compl	complementary	
conc	concentrate	
conc	concentrate [Canadian]	
conc	concentric	
conc.	concentrated [British]	
concd	concentrated	
concl	conclusion	
concn	concentration	
concn.	concentration [British]	
cond	condenser	
cond	condition	
cond	conditional	
cond	conductivity	
cond	conductivity [Canadian]	
conf	conference	
config	configuration	
congr	congruent	
conj	conjugate	
conj	conjunction	
conn	connect	
conn	connection	
consec	consecutive	
const	constant	
const	constant [Canadian]	
const.	constant [British]	
constr	construction	
cont	contact	
cont	continue	
cont	control	
contrib	contribution	
conv	convention	
conv	convergence	
conv	convert	
coord	coordinate	
coroll	corollary	
corr	correction	
corr	correspond	
corr	correspondence	
corr	corresponding	
correl	correlation	

cos	cosine	
cos	cosine [British]	
cos	cosine [Canadian]	
cos	cosine [ISO]	
cos	cosine [IUPAP]	
cos^{-1}	inverse cosine	
cos^{-1}	inverse cosine [British]	
cosec	cosecant [British]	
cosec	cosecant [ISO]	
cosec	cosecant [IUPAP]	
cosec^{-1}	inverse cosecant [British]	
cosech	hyperbolic cosecant [British]	
cosech	hyperbolic cosecant [ISO]	
cosech	hyperbolic cosecant [IUPAP]	
cosech^{-1}	inverse hyperbolic cosecant [British]	
cosh	hyperbolic cosine	
cosh	hyperbolic cosine [British]	
cosh	hyperbolic cosine [Canadian]	
cosh	hyperbolic cosine [ISO]	
cosh	hyperbolic cosine [IUPAP]	
cosh^{-1}	inverse hyperbolic cosine	
cosh^{-1}	inverse hyperbolic cosine [British]	
cot	cotangent	
cot	cotangent [British]	
cot	cotangent [Canadian]	
cot	cotangent [ISO]	
cot	cotangent [IUPAP]	
cot^{-1}	inverse cotangent	
cot^{-1}	inverse cotangent [British]	
coth	hyperbolic cotangent	
coth	hyperbolic cotangent [British]	
coth	hyperbolic cotangent [ISO]	
coth	hyperbolic cotangent [IUPAP]	
coth^{-1}	inverse hyperbolic cotangent	
coth^{-1}	inverse hyperbolic cotangent [British]	
counts/min	counts per minute	
counts/sec	counts per second	
covers	coversed sine	
c.p.	close-packed (crystals)	
cp	candlepower	
cp	candlepower [Canadian]	
cp	center of pressure	

cp	chemically pure
cp	chemically pure [Canadian]
cP	centipoise
cP	centipoise [British]
CPC	card-programmed calculator
cpd	contact potential difference
cpd.	compound [British]
c.p.h.	close-packed hexagonal (crystals)
cpl	couple
CPP	celestial principal point
cpr	scatter-specular ratio
cps	cycles per second
Cr	chromium
CR	cathode ray
CR	cold-rolled
CR	cooling rate
crit	criterion
crit	critical
crit.	critical [British]
crv	curve
cryog	cryogenics
cryst	crystal
c/s	cycles per second [British]
cS	centistoke
cS	centistoke [British]
Cs	cesium
CSA	celestial spin-axis point
c.&s.c.	capitals and small capitals
csc	cosecant
csc	cosecant [Canadian]
csc^{-1}	inverse cosecant
csch	hyperbolic cosecant
$csch^{-1}$	inverse hyperbolic cosecant
CSP	celestial subpoint
cstg	casting
CTC	coaxial thermal converter
ctg	cotangent [ISO]
ctg	cotangent [IUPAP]
ctgh	hyperbolic cotangent [IUPAP]
ctn	cotangent
ctn^{-1}	inverse cotangent
CTNF	controlled thermonuclear fusion

ctnh	hyperbolic cotangent	
ctnh⁻¹	inverse hyperbolic cotangent	
ctr	center	
cu	cubic	
cu	cubic [Canadian]	
Cu	copper (cuprum)	
cu cm	cubic centimeter [Canadian]	
cu ft	cubic foot [Canadian]	
cu in.	cubic inch [Canadian]	
cu m	cubic meter [Canadian]	
cum	cumulative	
cu mm	cubic millimeter [Canadian]	
cu mu	cubic micron [Canadian]	
cu μ	cubic micron [Canadian]	
CW	clockwise	
CW	continuous wave	
CX	critical experiment	
cyclo	cyclotron	
cyl	cylinder	
cyl	cylinder [Canadian]	
cyl	cylindrical	
Czech.	Czechoslovakian	
d	day [British]	
d	day [ISO]	
d	day [IUPAP]	
d	deci ($=10^{-1}$) (prefix) [International]	
d	deuteron	
D	debye	
D	deuterium	
D	down	
D/A	digital-to-analog	
Dan.	Danish	
db	decibel	
db	decibel [Canadian]	
dB	decibel [AIP]	
dB	decibel [British]	
dB	decibel [ISO]	
DB	dry bulb	
d.c.	direct current [British]	
d-c	direct-current (adjective)	
d-c	direct current (adjective only) [Canadian]	
dc	direct current	

DPMA

DC	decimal classification
DCX	direct-current experimental device
D-D	deuterium-deuterium reaction
DD	deuteron-deuteron (reactor)
DD	digital data
dec	decimal
dec	declination
Dec.	December
decr	decrease
ded	deduct
def	definite
def	definition
defl	deflect
defl	deflection
deg	degree
deg	degree [Canadian]
deg.	degree [British]
degC	degree Celsius (temperature interval) [British]
degF	degree Fahrenheit (temperature interval) [British]
degK	degree Kelvin (temperature interval) [British]
deion	deionizer
dele	delete
deliq	deliquescent
Dem	demonstration
denom	denominate
denom	denominator
dep	deposit
depr	depression
dept	department
deriv	derivation
deriv	derivative
des	design
descr	describe
desig	designation
det	determinant
det	determinant of
detm	determine
dev	develop
dev	deviate
dev	deviation
d.f.	degree of freedom
DF	decimal fraction

DF	distribution factor (radiation)
dg	decigram
D/H	deuterium-hydrogen ratio
dia	diameter [ASA]
diag	diagonal
diag	diagram
diam	diameter
diam	diameter [Canadian]
diaph	diaphragm
dict	dictionary
diff	difference
diff	differential
diffr	diffraction
dig comp	digital computer
dil	dilute
dil.	dilute [British]
dim.	dimension
dimorph	dimorphous
diop	diopter
dir	direct
dir	direction
dir	director
disc.	discrimination
disch	discharge
discrim	discriminant
disj	disjunction
displ	displacement
dis/sec	disintegrations per second [AIP]
dist	distance
distn	distortion
distr	distribution
div	divergence
div	divergence of
div	divide
div	division
dk	dark
dk	deka (=10) (prefix)
dkg	dekagram
dkl	dekaliter
dl	deciliter
DLF	dielectric loading factor
dm	decimeter

DM	detecting magnetometer
dm²	square decimeter
dm³	cubic decimeter
dn	delta amplitude, an elliptic function
dn	delta amplitude, an elliptic function [Canadian]
doc	document
dp	dew point
dps	disintegrations per second
dr	dram
dr	dram [Canadian]
dr	drive
dr	drum
DS	degree of substitution
DSP	disodium phosphate
DSV	dilute solution viscosity
D-T	deuterium-tritium reaction
DT	deuteron-triton (reactor)
DTA	differential thermoanalyzer
dtd	dated
dwg	drawing
dy	dyne [AAAS]
Dy	dysprosium
dyn	dynamic
dyn	dyne
dyn	dyne [British]
dyn	dyne [IUPAC]
dyn	dyne [IUPAP]
E	east
E	einsteinium
E	error
EAL	electromagnetic amplifying lens
EBOR	experimental beryllium oxide reactor
EBR	experimental breeder reactor
EBWR	experimental boiling-water reactor
EC	electron capture
ecc	eccentric
ech	echelon
econ	economist
ed	edition
ed.	editor
EDC	electronic digital computer
EDP	electronic data processing

EDPE	electronic data processing equipment
EDR	equivalent direct radiation
EE	electrical engineer
EF	exposure factor
eff	efficiency
eff	efficiency [Canadian]
e.g.	exempli gratia (for example)
EGCR	experimental gas-cooled reactor
EGO	eccentric orbiting geophysical observatory
ehf	extremely-high frequency
ehp	electric horsepower
Ei	exponential integral function
el	elevation
el	elevation [Canadian]
EL	elastic limit
EL	electroluminescence
elast	elasticity
elec	electric
elec	electric [Canadian]
elect.	electrolyte
elect.	electronics
electrochem	electrochemistry
electrophys	electrophysics
elem	element
elim	eliminate
ell.	ellipse
elong	elongation
EL-PC	electroluminescent-photoconductive
EL-PR	electroluminescent-photoresponsive
EM	electromagnetic
EM	electron microscope
e.m.f.	electromotive force [British]
emf	electromotive force
emf	electromotive force [Canadian]
emiss	emission
EMR	electromagnetic radiation
e.m.u.	electromagnetic unit [British]
emu	electromagnetic unit
emul	emulsion
ENE	east-northeast
eng	engine
engr	engineer

enum	enumeration	
env	envelope	
EOCR	experimental organic-cooled reactor	
EPR	electron paramagnetic resonance	
eq	equation	
eq	equation [Canadian]	
Eq.	equation [AIP]	
eqn.	equation [British]	
Eqs.	equations [AIP]	
equil	equilibrium	
equip.	equipment	
equiv	equivalent	
equiv.	equivalent [British]	
Er	erbium	
ERC	equatorial ring current	
erf	error function	
erf	error function of [British]	
erfc	error function, complementary	
es	electrostatic	
ESE	east-southeast	
ESR	electron spin resonance	
est	estimate	
estab	establishment	
e.s.u.	electrostatic unit [British]	
esu	electrostatic unit	
ET	ephemeris time	
et al.	et alibi (and elsewhere)	
et al.	et alii (and others)	
etc.	et cetera	
eth	ether	
et seq.	et sequens (and the following)	
eu	entropy unit	
Eu	europium	
ev	electron volt	
ev	evolves	
eV	electron volt [AIP]	
eV	electron volt [British]	
eV	electron volt [IUPAP]	
evap	evaporate	
evol	evolution	
E-W	east-west	
ex	example	

exam	examination	
exc	excitation	
exch	exchange	
exm	elementary transformation matrix	
exp	expand	
exp	expansion	
exp	exponential	
exp	exponential function of	
exp	exponential function of [British]	
exp	expulsion	
expl	explanation	
expos	exposure	
expr	expression	
expt	experiment	
expt.	experiment [British]	
expt.	experimental [British]	
exptl	experimental	
exsec	exsecant	
ext	extension	
ext	extent	
ext	exterior	
extr	extreme	
f	farad	
f	farad [Canadian]	
f	femto ($=10^{-15}$) (prefix) [International]	
f.	the following (in citations)	
F	degree Fahrenheit [ASA]	
F	degree Fahrenheit [Canadian]	
F	Fahrenheit	
F	failed	
F	false	
F	farad [AIP]	
F	farad [British]	
F	farad [IUPAC]	
F	farad [IUPAP]	
F	fermi	
F	fluorine	
F	force	
°F	degree Fahrenheit	
°F	degree Fahrenheit [IUPAC]	
°F	degree Fahrenheit (temperature value) [British]	
F°	Fahrenheit degree (temperature difference) [AAAS]	

fall.	fallacy	
fbr	fiber	
FBR	fast-breeder reactor	
f.c.	face-centered (crystals)	
f.c.c.	face-centered cubic (crystals)	
fcc	face-centered cubic (crystals) [AIP]	
fd	farad [AAAS]	
Fe	iron (ferrum)	
Feb.	February	
ff.	the following (plural) (in citations)	
fhp	fractional horsepower	
fig.	figure [British]	
Fig.	Figure	
fil	filament	
Finn.	Finnish	
fl	floor	
fl	fluid	
fl	fluid [Canadian]	
fl dr	fluid dram	
FLIP	floating index point	
FLIP	floating laboratory instrument platform	
fl oz	fluid ounce	
flt	float	
FLT	force-length-time system	
fluor	fluorescent	
Fm	fermium	
FM	frequency modulation	
FMLT	force-mass-length-time system	
fnp	fusion point	
fnp	fusion point [Canadian]	
form.	formula	
f.p.	freezing point [British]	
fp	freezing point	
fp	freezing point [Canadian]	
F.P.	fission product [AEC]	
FP	full period	
FPIS	forward propagation by ionospheric scatter	
fps	feet per second	
fps	feet per second [Canadian]	
fps	foot-pound-second	
fps	foot-pound-second [Canadian]	
FPS	Fabry-Perot spherical (interferometer)	

fpsu	foot-pound-second unit	
FPTS	forward propagation by tropospheric scatter	
fr	fresnel	
Fr	francium	
Fr	franklin	
Fr	frontier set of	
Fr.	French	
frac	fraction	
frac	fractional	
freq	frequency	
Fri.	Friday	
fs	fine structure	
FS	Fourier series	
ft	foot	
ft	foot [British]	
ft	foot [Canadian]	
ft	foot [ISO]	
ft²	square foot	
ft²	square foot [British]	
ft²	square foot [ISO]	
ft³	cubic foot	
ft³	cubic foot [British]	
ft³	cubic foot [ISO]	
ft-c	foot-candle	
ft-c	foot-candle [Canadian]	
ft-L	foot-lambert	
ft-L	foot-lambert [Canadian]	
ft-lam	foot-lambert [AAAS]	
ft-lb	foot-pound	
ft-lb	foot-pound [Canadian]	
ft lbf	foot-pound-force [British]	
ft-lbf	foot-pound-force	
ft/s	feet per second [ISO]	
ft/s²	feet per second squared [ISO]	
ful	fulcrum	
fwd	forward	
FWHM	full width at half maximum	
g	gram	
g	gram [British]	
g	gram [Canadian]	
g	gram [IUPAC]	
g	gram [IUPAP]	

DPMA

g	gravity	
g	grid	
G	gauss	
G	gauss [British]	
G	gauss [IUPAP]	
G	giga (=10^9) (prefix) [International]	
ga	gage	
Ga	gallium	
gal	gallon	
gal	gallon [British]	
gal	gallon [Canadian]	
Gal	gal	
Gal	gal [ISO]	
galv	galvanometer	
GAT	Greenwich apparent time	
Gc	gigacycle	
g/cc	grams per cubic centimeter	
gcd	greatest common divisor	
gcd	greatest common divisor [Canadian]	
G.C.D.	greatest common divisor	
gcf	greatest common factor	
G.C.F.	greatest common factor	
g/cm³	grams per cubic centimeter	
GCR	gas-cooled reactor	
gcs	greatest common subgroup	
G.C.S.	greatest common subgroup	
gd	Gudermannian	
Gd	gadolinium	
Ge	germanium	
GEK	geomagnetic electrokinetograph	
gen	general	
gen	generator	
geochem	geochemistry	
geod	geodetic	
geol	geological	
geom	geometric	
geom	geometry	
geomag	geomagnetism	
geophys	geophysics	
Ger.	German	
Gev	gigaelectron volt	
GeV	gigaelectron volt [LRL]	

gf	gram-force	
GF	Galois field	
GHA	Greenwich hour angle	
GHz	gigahertz	
Gi	gilbert	
g.l.b.	greatest lower bound	
GLIPAR	Guide Line Identification Program for Antimissile Research	
glob	globular	
gloss.	glossary	
g-m	gram-meter	
gm	gram-mass	
GM	Geiger-Muller (radiation counter)	
GM	geometric mean	
GMST	Greenwich mean sidereal time	
GMT	Greenwich mean time	
gmv	gram-molecular volume	
gp	group	
GP	geometric progression	
gpm	geopotential meter	
gr	grade	
gr	grain	
gr	grain [British]	
grad	gradient	
grad	gradient of	
gran	granular	
grtg	grating	
GST	Greenwich sidereal time	
G-T	Gamow-Teller (selection rules for beta decay)	
GYS	guaranteed yield strength	
h	hecto ($=10^2$) (prefix)	
h	henry [Canadian]	
h	hour	
h	hour [British]	
h	hour [ISO]	
h	hour [IUPAC]	
h	hour [IUPAP]	
h	hour (in astronomical tables)	
h	hour (in astronomical tables) [Canadian]	
h.	hot	
H	henry [AIP]	
H	henry [British]	

H	henry [IUPAC]	
H	henry [IUPAP]	
H	hydrogen	
HA	hour angle	
HADEC	hour angle declination	
hav	haversine	
hav	haversine [Canadian]	
hcd	high current density	
hcf	highest common factor	
H.C.F.	highest common factor	
h.c.p.	hexagonal close-packed (crystals)	
hcp	hexagonal close-packed (crystals) [AIP]	
hdbk	handbook	
He	helium	
heterog	heterogeneous	
hex	hexagon	
hf	high frequency	
Hf	hafnium	
hfs	hyperfine structure	
hg	hectogram	
Hg	mercury	
H.I.	heat input	
HILAC	heavy ion linear accelerator	
hl	hectoliter	
HM	harmonic mean	
Ho	holmium	
homo	homogeneous	
hon.	honorary	
hor	horizon	
hp	high pressure	
hp	horsepower	
hp	horsepower [British]	
hp	horsepower [Canadian]	
HP	harmonic progression	
hr	hour [Canadian]	
HR	hot-rolled	
ht	heat	
HTGCR	high-temperature gas-cooled reactor	
HTGR	high-temperature gas reactor	
HTS	high tensile strength	
HTV	hypersonic test vehicle	
HWCTR	heavy water components test reactor	

HWGCR	heavy-water-moderated gas-cooled reactor
HWR	hot-water reactor
hyd	hydraulic
hyd	hydrolyzed
hydrodyn	hydrodynamic
hydroelec	hydroelectric
hydrol	hydrology
hydromech	hydromechanical
hyg	hygroscopic
hyp	hypotenuse
hyp	hypothesis
hyperb	hyperbola
hyperbol	hyperbolic
hz	hertz [AAAS]
Hz	hertz
Hz	hertz [ISO]
Hz	hertz [IUPAC]
Hz	hertz [IUPAP]
I	iodine
I.A.	inscribed angle
Iλ	international angstrom
IAD	International Astrophysical Decade
IAGS	Inter-American Geodetic Survey
IBC	international brightness coefficient
ibid.	ibidem (in the same place)
I.C.	inscribed circle
IC	ion chamber
ICEF	International Cooperative Emulsion Flight (study)
ICT	International Critical Tables
i.d.	inside diameter
id.	idem (the same)
ID	isotope dilution
ident	identical
i.e.	id est (that is)
IE	impact energy
IE	index error
i.f.	intermediate frequency
IF	internal friction
IFT	interfacial tension
IGC	International Geophysical Cooperation
IGY	International Geophysical Year
illus	illustration

Im	imaginary part of	
Im	imaginary part of [British]	
Im	imaginary part of [ISO]	
Im	imaginary part of [IUPAP]	
IMA	ideal mechanical advantage	
imag	imaginary	
imp.	impact	
imp.	impulse	
IMW	International Map of the World on the Millionth Scale	
in	inch [British]	
in	inch [ISO]	
in.	inch	
in.	inch [Canadian]	
In	indium	
in^2	square inch [British]	
in^2	square inch [ISO]	
$in.^2$	square inch	
in^3	cubic inch [British]	
in^3	cubic inch [ISO]	
$in.^3$	cubic inch	
incl	inclination	
incl	including	
incl	inclusive	
incoh	incoherent	
incr	increase	
incr	increment	
ind	inductance	
ind	induction	
ind	industry	
indef	indefinite	
indep	independent	
indet	indeterminate	
indiv	individual	
ineq	inequality	
inf	infinity	
inf.	infinum	
infer.	inference	
infl	inflection	
inflam	inflammable	
info	information	
in. Hg	inch of mercury	

inHg	inch of mercury [British]	
init	initial	
in.-lb	inch-pound [LRL]	
in-lb	inch-pound [ASA]	
in-lb	inch-pound [Canadian]	
i.n. mi.	international nautical mile	
in. oil	inch of oil	
inorg	inorganic	
insol	insoluble	
inst	institute	
instr	instructor	
int	integer	
int	integral	
int	integrate	
int	intercept	
int	interest	
int	interior	
int	internal	
int	internal [Canadian]	
int	intersect	
int	interval	
interp	interpolation	
intl	international	
inv	inverse	
invar	invariant	
invol	involute	
inx	index	
IPP	image principal point	
ips	inches per second	
ips	inches per second [Canadian]	
IPY	International Polar Year	
i.q.	idem quod (the same as)	
IQSY	International Year of the Quiet Sun	
i.r.	infrared [British]	
i.r.	inside radius	
ir	infrared	
Ir	iridium	
iraser	infrared amplification by stimulated emission of radiation	
IRE	ion rocket engine	
IRMP	infrared measurement program	
irrad	irradiation	

IS	impact strength
ISA	image spin-axis point
ISIS	International Satellites for Ionospheric Studies (program)
isol	isolate
isom	isometric
isos	isosceles
isoth	isothermal
IT	isomeric transition
Ital.	Italian
IU	international unit
ivp	initial vapor pressure
IWDS	International World Day Service
j	joule [Canadian]
J	joule
J	joule [AIP]
J	joule [British]
J	joule [IUPAC]
J	joule [IUPAP]
Jan.	January
Japan.	Japanese
k	kilo (=10^3) (prefix) [International]
K	degree Kelvin [ASA]
K	degree Kelvin [Canadian]
K	kaiser
K	Kelvin
K	potassium
°K	degree Kelvin
°K	degree Kelvin [ISO]
°K	degree Kelvin [IUPAC]
°K	degree Kelvin [IUPAP]
°K	degree Kelvin (temperature value) [British]
K°	Kelvin degree (temperature difference) [AAAS]
Kans.	Kansas
kb	kilobar
kc	kilocurie
kc	kilocycle
kc	kilocycles per second [Canadian]
kC	kilocurie [LRL]
kcal	kilocalorie
kcal	kilocalorie [British]
kcal	kilocalorie [Canadian]

kcal	kilocalorie [IUPAP]
kcal/mole	kilocalories per mole
kc/sec	kilocycles per second [AIP]
KE	**kinetic energy**
kerma	kinetic energy released in material
kev	kiloelectron volt
keV	kiloelectron volt [AIP]
kg	kilogram
kg	kilogram [British]
kg	kilogram [Canadian]
kg	kilogram [ISO]
kg	kilogram [IUPAP]
kG	kilogauss
kg/cm^2	kilograms per square centimeter
kgf	kilogram-force
kgf	kilogram-force [British]
kg-m	kilogram-meter
kg-m	kilogram-meter [Canadian]
kgm	kilogram-mass
kg/m^3	kilograms per cubic meter
kg/m^3	kilograms per cubic meter [Canadian]
kg-mole	kilogram-mole
kg per cu m	kilograms per cubic meter [Canadian]
kgps	kilograms per second [Canadian]
kg/sec	kilograms per second
kg-wt	kilogram-weight
kHz	kilohertz
kin.	kinetic
kips	kilopounds
kj	kilojoule
kJ	kilojoule [AIP]
kl	kiloliter
kl	kiloliter [Canadian]
kliter	kiloliter [LRL]
km	kilometer
km	kilometer [Canadian]
kM	kilomega (=10^9) (prefix, giga preferred)
kOe	kilo-oersted
kpc	kiloparsec
Kr	krypton
kv	kilovolt
kv	kilovolt [Canadian]

kV	kilovolt [AIP]
kva	kilovolt-ampere [Canadian]
kVA	kilovolt-ampere [AIP]
kvar	reactive kilovolt-ampere [Canadian]
kvar	reactive kilovolt-ampere; kilovar
kw	kilowatt
kw	kilowatt [Canadian]
kW	kilowatt [AIP]
kW	kilowatt [British]
kWh	kilowatt-hour [AIP]
kWh	kilowatt-hour [British]
kWh	kilowatt-hour [IUPAP]
kwhr	kilowatt-hour [Canadian]
KWIC	key word in context (indexing system)
kΩ	kilohm
l	line
l	liter
l	liter [British]
l	liter [Canadian]
l	liter [ISO]
l	liter [IUPAČ]
l	liter [IUPAP]
l	lumen [Canadian]
L	lambert
L	lambert [Canadian]
La	lanthanum
LAAR	liquid air accumulator rocket
lab	laboratory
laby	labyrinth
lam	lambert [AAAS]
laser	light amplification by stimulated emission of radiation
lat	lateral
lat	latitude
lat	latitude [Canadian]
LAT	local apparent time
lat ht	latent heat
lat.ht.	latent heat [British]
lb	binary logarithm [ISO]
lb	binary logarithm [IUPAP]
lb	pound
lb	pound [British]

lb	pound [Canadian]
lb avdp	pound, avoirdupois
lbf	pound-force
lbf	pound-force [British]
lbf-ft	pound-force-foot
lbf/in.²	pound-force per square inch
lb-ft	pound-foot
lb-ft	pound-foot [Canadian]
lb-in.	pound-inch
lb-in.	pound-inch [Canadian]
lbm	pound-mass
lb per cu ft	pounds per cubic foot [ASA]
lb per cu ft	pounds per cubic foot [Canadian]
lb t	pound, troy
l.c.	lower case
LC	inductance-capacitance
lcb	longitudinal center of buoyancy
lcd	lowest common denominator
L.C.D.	lowest common denominator
lcf	least common factor
lcf	longitudinal center of flotation
L.C.F.	least common factor
lcg	longitudinal center of gravity
lcm	least common multiple
lcm	least common multiple [Canadian]
lcm	lowest common multiple
L.C.M.	least common multiple
L.C.M.	lowest common multiple
LCR	inductance-capacitance-resistance
LET	linear energy transfer
lf	low frequency
lg	common logarithm [ISO]
lg	decadic logarithm [IUPAP]
LHA	local hour angle
l-hr	lumen-hour [Canadian]
li	logarithmic integral function
Li	lithium
l.i.m.	limit-in-mean
lim	limit
lim	limit of [British]
lim	limit of [ISO]
lin	linear

LINAC	linear accelerator	
lin eq	linear equation	
liq	liquid	
liq	liquid [Canadian]	
liq.	liquid [British]	
lit.	literature	
lkg	leakage	
ll	lines	
lm	lumen	
lm	lumen [British]	
lm	lumen [IUPAC]	
lm	lumen [IUPAP]	
lm-hr	lumen-hour	
lm/m²	lumens per square meter	
lm-sec	lumen-second	
LMT	local mean time	
LMTD	logarithmic mean temperature difference	
lm/W	lumens per watt [AIP]	
ln	natural logarithm	
ln	natural logarithm [British]	
ln	natural logarithm [Canadian]	
ln	natural logarithm [ISO]	
ln	natural logarithm [IUPAP]	
loc	locate	
loc	locus	
loc. cit.	loco citato (in the place cited)	
log	common logarithm [ISO]	
log	common logarithm [British]	
log	decadic logarithm [IUPAP]	
log	logarithm	
log.	logic	
log₂	binary logarithm [ISO]	
log₂	binary logarithm [IUPAP]	
log₁₀	common logarithm [British]	
log₁₀	common logarithm [ISO]	
log₁₀	logarithm to base 10	
logₐ	logarithm to the base a [ISO]	
logₐ	logarithm to the base a [IUPAP]	
logₑ	natural logarithm	
logₑ	natural logarithm [British]	
logₑ	natural logarithm [Canadian]	
logₑ	natural logarithm [ISO]	

long.	longitude	
long.	longitude [Canadian]	
LOS	line of sight	
lp	low pressure	
LP	linear programming	
lpw	lumens per watt [ASA]	
lpw	lumens per watt [Canadian]	
lpW	lumens per watt [IEC]	
LRG	long range	
LSR	linear straining rate	
LST	local sidereal time	
lt	light	
L.T.E.	local thermodynamics equilibrium	
lt-yr	light-year	
Lu	lutetium	
l.u.b.	least upper bound	
lvr	lever	
Lw	lawrencium	
lx	lux	
lx	lux [British]	
lx	lux [IUPAC]	
lx	lux [IUPAP]	
ly	langley	
LYP	lower yield point	
LZT	local zone time	
m	meter	
m	meter [British]	
m	meter [Canadian]	
m	meter [ISO]	
m	meter [IUPAC]	
m	meter [IUPAP]	
m	milli ($=10^{-3}$) (prefix) [International]	
m	minute (time, in astronomical tables)	
m	minute (time, in astronomical tables) [Canadian]	
M	molar	
M	mega ($=10^6$) (prefix) [International]	
M	molar (concentration) [British]	
M	mole [LRL]	
M	thousand	
M	thousand [Canadian]	
M%	mole percent	
m^{-1}	reciprocal meter [ISO]	

m²	square meter	
m²	square meter [British]	
m²	square meter [Canadian]	
m²	square meter [ISO]	
m³	cubic meter	
m³	cubic meter [British]	
m³	cubic meter [Canadian]	
m³	cubic meter [ISO]	
ma	milliampere	
ma	milliampere [Canadian]	
mA	milliampere [AIP]	
mÅ	milliangstrom	
mach	machine	
mag	magnet	
mag	magnitude	
maj	major	
maj ax.	major axis	
man.	manual	
manom	manometer	
Mar.	March	
maser	microwave amplification by stimulated emission of radiation	
mat.	matrix	
math	mathematical	
math	mathematics	
max	maximum	
max	maximum [Canadian]	
max.	maximum [British]	
mb	millibar	
mb	millibar [British]	
mb	millibarn	
mc	millicurie	
mC	millicurie [LRL]	
Mc	megacurie	
Mc	megacycle	
MC	megacurie [LRL]	
mCi	millicurie [AIP]	
Mc/sec	megacycles per second [AIP]	
MD	mean deviation	
ME	mechanical engineer	
meas	measure	
mech	mechanical	

mechs	mechanics	
med	median	
med	medium	
MEE	margin of elastic energy	
MEP	Moon-Earth plane	
meq.	milliequivalent [ACS]	
met.	metallurgical	
metab	metabolism	
Mev	million electron volts	
MeV	million electron volts [AIP]	
mf	medium frequency	
MFI	melt flow index	
mfp	mean free path	
MFP	mixed fission products	
m.g.	motor generator [LRL]	
mg	milligram	
mg	milligram [Canadian]	
Mg	magnesium	
mGal	milligal	
MGD	magnetogasdynamics	
mh	millihenry [Canadian]	
mH	millihenry [AIP]	
mhc	mean horizontal candle	
mhcp	mean horizontal candlepower	
mhcp	mean horizontal candlepower [Canadian]	
MHD	magnetohydrodynamics	
mic	micrometer	
micr	microscope	
midpt	midpoint	
mil	military	
mile/h	miles per hour [ISO]	
min	mineral	
min	minimum	
min	minimum [Canadian]	
min	minute	
min	minute [Canadian]	
min	minute [ISO]	
min	minute [IUPAC]	
min	minute [IUPAP]	
min	minute (time) [British]	
min.	minimum [British]	
min^{-1}	reciprocal minute [ISO]	

min ax.	minor axis
mineral.	mineralogy
misc	miscellaneous
misc	miscible
MIST	multigroup internuclear slab transport
mix.	mixture
mk	mark
m-kg	meter-kilogram
m-kg	meter-kilogram [Canadian]
mks	meter-kilogram-second
mksa	meter-kilogram-second-ampere
mksm	meter-kilogram-second electromagnetic unit
mksu	meter-kilogram-second unit
ml	milliliter
ml	milliliter [British]
ml	milliliter [Canadian]
mL	millilambert
mL	millilambert [Canadian]
mliter	milliliter [LRL]
MLT	mass-length-time system
mm	millimeter
mm	millimeter [British]
mm	millimeter [Canadian]
mM	millimole
Mm	midmean
MM	magnetic moment
MM	megamega ($=10^{12}$) (prefix, tera preferred)
mm^2	square millimeter
mm^2	square millimeter [Canadian]
mm^3	cubic millimeter
mm^3	cubic millimeter [Canadian]
m.m.f.	magnetomotive force [British]
mmf	magnetomotive force
mm Hg	millimeter of mercury
mmHg	millimeter of mercury [British]
m mu	millimicron [Canadian]
mmu	milli-mass-units
mMU	milli-mass-units [LRL]
Mn	manganese
MN	magnetic north
mo	month
Mo	mode

Mo	molybdenum
mod	model
mod	moderate
mod	modulo
mod	modulus
Mohm	megohm [AAAS]
mol	mole [IUPAP]
mol	molecular
mol	molecule
mol.	molecular [British]
mol.	molecule [British]
mole	gram-molecule [British]
mol wt	molecular weight
mol. wt	molecular weight [ASA]
mol. wt	molecular weight [Canadian]
mol. wt.	molecular weight [British]
Mon.	Monday
monog	monograph
mot	motor
m.p.	melting point [British]
mp	melting point
mp	melting point [Canadian]
MPCD	minimum perceptible color difference
mph	miles per hour
mph	miles per hour [Canadian]
MPRE	minimum pure radium equivalent
mr	milliroentgen
MR	medium range
MRD	method of rapid determination
mrem	millirem
mr/hr	milliroentgens per hour
MRN	Meteorological Rocket Network
mr/yr	milliroentgens per year
m/s	meters per second [British, I.E.E.]
m/s	meters per second [ISO]
MS	magnetic south
MS	mean square
m/s^2	meters per second squared [ISO]
msc	mean spherical candle
mscp	mean spherical candlepower
MSD	mean square difference
MSE	mean square error

m/sec	meters per second	
msec	millisecond	
msl	missile	
MSL	mean sea level	
mt	mount	
Mt	megaton [LRL]	
MT	megaton	
mtg	mounting	
mthd	method	
mu	micron [Canadian]	
mu a	microampere [ASA]	
mu a	microampere [Canadian]	
mult	multiple	
mult	multiplication	
mu mu	micromicron [Canadian]	
muon	mu-meson	
mu w	microwatt [Canadian]	
mv	millivolt	
mv	millivolt [Canadian]	
mV	millivolt [AIP]	
Mv	mendelevium	
MV	mean variation	
MV	million volts [AIP]	
mw	milliwatt	
mW	milliwatt [AIP]	
Mw	megawatt	
MW	megawatt [LRL]	
MW	microwave	
Mx	maxwell	
Mx	maxwell [IUPAP]	
my	myria ($=10^4$) (prefix)	
myg	myriagram	
myl	myrialiter	
mzcp	mean zonal candlepower	
mμ	millimicro ($=10^{-9}$) (prefix, nano preferred)	
mμ	millimicron	
mμ	millimicron [Canadian]	
mμsec	millimicrosecond	
MΩ	megohm	
n	nano ($=10^{-9}$) (prefix) [International]	
n.	noun	
N	nadir	

N	neper [British]	
N	newton	
N	newton [British]	
N	newton [IUPAC]	
N	newton [IUPAP]	
N	nitrogen	
N	normal (concentration)	
N	normal (concentration) [British]	
N	normal (concentration) [IUPAC]	
N	north	
Na	sodium	
N.A.	numerical aperture	
nat	natural	
Nb	niobium (columbium)	
N.B.	nota bene (mark well)	
nc	nanocurie	
Nd	neodymium	
Ne	neon	
NE	northeast	
neg	negative	
net.	network	
neut	neutral	
Ni	nickel	
n.m.	nuclear magneton	
nm	nanometer	
NMR	nuclear magnetic resonance	
NNE	north-northeast	
NNES	National Nuclear Energy Series (of AEC)	
NNW	north-northwest	
no.	number	
No	nobelium	
No.	numero (number)	
NOMSS	National Operational Meterological Satellite System	
N-on-P	negative-on-positive (solar cell)	
norm.	normal	
Nov.	November	
Np	neper	
Np	neper [ISO]	
Np	neptunium	
NP	nonparametric	
NPD	nuclear power demonstration reactor	

DPMA

NR	noise reduction
NRC	noise reduction coefficient
NRS	nonreflected-shock tunnel
ns	nanosecond [IUPAP]
N-S	north-south
NSA	Nuclear Science Abstracts (of AEC)
nsec	nanosecond
NSRDS	National Standard Reference Data System
nt	nit
nt	nit [British]
NTP	normal temperature and pressure
nucl	nuclear
nudet	nuclear detonation
num	numeral
num	numerator
nvt	integrated neutron flux
NW	northwest
O	oxygen
OAO	orbiting astronomical observatory
obj	object
obj	objective
obs	observation
obs	observed
obs.	observed [British]
obt	obtuse
obv	obverse
oceanog	oceanography
oct	octagon
oct	octal
Oct.	October
octahdr	octahedral
o.d.	outside diameter
Oe	oersted
Oe	oersted [IUPAP]
oer	oersted [AAAS]
OGO	orbiting geophysical observatory
ohm.	ohmmeter
OMR	organic-moderated reactor
op. cit.	opere citato (in the work cited)
oper	operator
OPO	orbiting planetary observatory
opp	opposite

opt	optics
o.r.	outside radius
orb.	orbit
ord	ordinal
ord	ordinary
org	organic
orient.	orientation
orig	origin
orthog	orthogonal
Os	osmium
osc	oscillate
OSO	orbiting solar observatory
OA	over-all
ovv	overvoltage
oz	ounce
oz	ounce [British]
oz	ounce [Canadian]
oz avdp	ounce, avoirdupois
ozf	ounce-force [British]
oz-ft	ounce-foot
oz-ft	ounce-foot [Canadian]
oz-in.	ounce-inch
oz-in.	ounce-inch [Canadian]
oz t	ounce, troy
p	pico ($=10^{-12}$) (prefix) [International]
p.	page
P	passed
P	phosphorus
P	poise
P	poise [British]
P	poise [IUPAC]
P	poise [IUPAP]
Pa	protactinium
PAM	pulse-amplitude modulation
par.	paragraph
par.	parallel
parab	parabola
param	parameter
param	parametric
parens	parentheses
parsec	parallax second
part.	participle

part.	particle	
Pb	lead	
pc	parsec	
pc	picocurie	
PCA	point of closest approach	
PCA	polar-cap absorption	
pcf	pounds per cubic foot	
PCM	pulse-code modulation	
PCM	pulse-count modulation	
pct	percent	
p.d.	potential difference [British]	
pd	period	
pd	prism diopter	
Pd	palladium	
P.D.	principal distance	
PD	potential difference	
pdl	poundal	
pdl	poundal [British]	
PDM	pulse-duration modulation	
PDR	power demonstration reactor	
pe	probable error [AIP]	
PE	probable error	
pent.	pentagon	
perf	perfect	
perm	permutation	
perp	perpendicular	
perp bis	perpendicular bisector	
persp	perspective	
pf	picofarad	
pF	picofarad [AIP]	
pF	picofarad [IUPAP]	
PFM	pulse-frequency modulation	
PGF	pressure gradient force	
p-gon	polygon	
ph	phase	
ph	phot	
pH	hydrogen-ion concentration	
pH	hydrogen-ion concentration [British]	
p-hed	polyhedron	
phys	physical	
phys	physics	
p.i.	point of inflection	

PIG	Penning ionization gage	
PIG	Philips ionization gage	
PILAC	pulsed ion linear accelerator	
pion	pi-meson	
PIQSY	probes for the International Quiet Solar Year	
pl	plane	
pl	plate	
pl.	plural	
PLA	proton linear accelerator	
pln	plenum	
p.m.	post meridiem (afternoon)	
pm	permanent magnet	
Pm	promethium	
PM	pulse modulation	
p-nom	polynomial	
Po	polonium	
POGO	polar orbiting geophysical observatory	
pol	polarity	
Pol.	Polish	
P-on-N	positive-on-negative (solar cell)	
POPR	prototype organic power reactor	
pos	positive	
post.	postulate	
pot.	potentiometer	
pp.	pages	
ppb	parts per billion	
ppm	parts per million	
ppm	parts per million [Canadian]	
ppm	pulses per minute	
PPM	pulse-position modulation	
pps	pulses per second	
ppt	precipitate	
ppt.	precipitate [British]	
pptn	proportion	
pptnl	proportional	
pr	pair	
pr	prisms	
Pr	praseodymium	
prep.	preposition	
press.	pressure	
prob	probability	
prob	problem	

proc	process	
prod.	product	
Prof.	Professor	
prog	program	
proj	project	
proj	projection	
pron.	pronoun	
prop.	proposition	
pro tem.	pro tempore (temporarily)	
prt	platinum resistance thermometer	
PRTR	plutonium recycle test reactor	
PS	power series	
PS	proton synchrotron	
psec	picosecond	
psf	pounds per square foot	
psf	pounds per square foot [Canadian]	
psi	pounds per square inch	
psi	pounds per square inch [Canadian]	
pt	part	
pt	pint	
pt	pint [Canadian]	
pt	point	
Pt	platinum	
Pt-Co	platinum-cobalt (color method)	
PTM	pulse-time modulation	
PTR	pool test reactor	
Pu	plutonium	
publ	publication	
pul	pulley	
pvp	partial vapor pressure	
P-wave	pressure wave	
PWM	pulse-width modulation	
PWR	pressurized water reactor	
pyr	pyramid	
pyr	pyrometer	
QD	quartile deviation	
Q.E.D.	quod erat demonstrandum (which was to be demonstrated, or proved)	
Q.E.F.	quod erat faciendum (which was to be done)	
QF	quality factor (radiation)	
QL	quasi-longitudinal	
qq.v.	quae vide (which see) (plural)	

qt	quart
qt	quart [Canadian]
QT	quasi-transverse
qtr	quarter
quad	quadrangle
quad	quadrant
quad	quadratic
quad	quadrilateral
quant	quantum
quant mech	quantum mechanics
quant no.	quantum number
QUASER	quantum amplification by stimulated emission of radiation
quot	quotient
q.v.	quod vide (which see)
Q-wave	love wave
r	ratio
r	roentgen
r	roentgen [British]
R	range
R	resistor
R	roentgen [AIP]
Ra	radium
RA	right ascension
rad	radian
rad	radian [British]
rad	radian [ISO]
rad	radiant
rad	radical
rad	radius
radar	radio detection and ranging
radiog	radiography
radn	radiation
rad/s	radians per second [British]
rad/s	radians per second [ISO]
RAMS	right ascension of mean sun
ratnl	rational
Rb	rubidium
RBE	relative biological effectiveness
RC	reactivity coefficient
RC	resistance-capacitance
RCG	radioactivity concentration guides

R&D	research and development	
re	regarding	
Re	real part of	
Re	real part of [British]	
Re	real part of [ISO]	
Re	real part of [IUPAP]	
Re	rhenium	
reac	reactivity	
reac	reactor	
recap	recapitulation	
recip	reciprocal	
rect	rectangle	
red.	reduce	
redox	reduction-oxidation	
ref	reference	
refl	reflection	
refr	refraction	
reg	region	
reg	regular	
rel	relative	
rem	remainder	
rem	roentgen equivalent, man	
rep	repulsion	
rep	roentgen equivalent, physical	
repro	reproduction	
rept	report	
res	research	
res	resistor	
resid	residual	
resp	respectively	
rev	reverse	
rev	review	
rev	revolution	
rev/min	revolutions per minute [British]	
rf	radio-frequency (adjective)	
Rh	rhodium	
Rh	Rockwell hardness	
r/hr	roentgens per hour	
RI	refractive index	
riometer	relative ionospheric opacity meter	
rkt	rocket	
RLE	rate of loss of energy	

RMF	reactivity measurement facility	
r.m.s.	root-mean-square [British]	
rms	root-mean-square	
rms	root-mean-square [Canadian]	
rmu	rest mass unit	
Rn	radon	
rot	rotation	
rot.	rotate	
ROT	rule of thumb	
Roy.	Royal	
rpm	revolutions per minute	
rpm	revolutions per minute [Canadian]	
rps	revolutions per second	
rps	revolutions per second [Canadian]	
rss	root-sum-square	
rt	right	
rt ang	right angle	
RTC	radiation-thermal cracking	
Ru	ruthenium	
Russ.	Russian	
r.v.	random variable	
R-wave	Rayleigh wave	
Ry	rydberg	
s	second [ISO]	
s	second [IUPAC]	
s	second [IUPAP]	
s	second (time) [British]	
s	second (time, in astronomical tables)	
s	second (time, in astronomical tables) [Canadian]	
S	south	
S	stoke	
S	stoke [British]	
S	sulfur	
s^{-1}	reciprocal second [ISO]	
s^{-2}	reciprocal second squared [ISO]	
SAI	sudden auroral intensity	
samp	sample	
SANE	scientific applications of nuclear explosions	
SAO	Smithsonian Astrophysical Observatory	
s.a.s.	side, angle, side	
sat.	saturate	
Sat.	Saturday	

satel	satellite
satn	saturation
sb	stilb
sb	stilb [British]
Sb	antimony
sc	scales
Sc	scandium
SCF	self-consistent field (calculations)
sch	schedule
sci	science
sci	scientist
SCM	symmetrically cyclically magnetized (condition)
SCNA	sudden cosmic noise absorption
scp	spherical candlepower
scp	spherical candlepower [Canadian]
sd	semidiameter
SD	standard deviation
Se	selenium
SE	southeast
SE	standard error
SEA	sudden enhancement of atmospherics
sec	secant
sec	secant [British]
sec	secant [Canadian]
sec	secant [ISO]
sec	secant [IUPAP]
sec	second
sec	second [Canadian]
Sec.	Section (in references)
sec^{-1}	inverse secant
sec^{-1}	inverse secant [British]
sech	hyperbolic secant
sech	hyperbolic secant [British]
sech	hyperbolic secant [ISO]
sech	hyperbolic secant [IUPAP]
$sech^{-1}$	inverse hyperbolic secant
$sech^{-1}$	inverse hyperbolic secant [British]
sect.	section
sect.	sector
seg	segment
Sept.	September
seq	sequence

SET	Solar Energy Thermionic (program)
SETS	Solar Energy Thermionic Conversion System
SGR	sodium graphite reactor
SHA	sidereal hour angle
shf	super-high frequency
SHM	simple harmonic motion
si	second sine-integral function
Si	first sine-integral function
Si	silicon
sic	specific inductive capacity
SID	sudden ionospheric disturbance
sim	similar
simul	simultaneous
simul eqs	simultaneous equations
sin	sine
sin	sine [British]
sin	sine [Canadian]
sin	sine [ISO]
sin	sine [IUPAP]
sin^{-1}	inverse sine
sin^{-1}	inverse sine [British]
sinh	hyperbolic sine
sinh	hyperbolic sine [British]
sinh	hyperbolic sine [Canadian]
sinh	hyperbolic sine [ISO]
sinh	hyperbolic sine [IUPAP]
$sinh^{-1}$	inverse hyperbolic sine
$sinh^{-1}$	inverse hyperbolic sine [British]
Sm	samarium
sn	sine of the amplitude, an elliptic function
sn	sine of the amplitude, an elliptic function [Canadian]
Sn	tin (stannum)
S-N curve	stress to number of cycles curve
snd	sound
snr	signal-noise ratio [AGU]
SNR	signal-noise ratio
sol	solar
sol	solid
solv	solvent
sp act.	specific activity
Span.	Spanish

spec	specification
sp gr	specific gravity
sp gr	specific gravity [Canadian]
sp.gr.	specific gravity [British]
sph	sphere
spher	spherical
sp ht	specific heat
sp ht	specific heat [Canadian]
sp.ht.	specific heat [British]
SPL	sound pressure level
SPM	stochastic process method
sp vol	specific volume
sq	square
sq	square [Canadian]
sq cm	square centimeter
sq cm	square centimeter [Canadian]
sq ft	square foot [Canadian]
sq in.	square inch [Canadian]
sq m	square meter [Canadian]
sq mm	square millimeter [Canadian]
sq mu	square micron [Canadian]
sq rt	square root
sq μ	square micron [Canadian]
sr	steradian
sr	steradian [ISO]
Sr	strontium
SR	short range
SRF	self-resonant frequency
SS	sum of squares
SSE	south-southeast
s.s.s.	side, side, side
SSW	south-southwest
S-SWF	sudden short-wave fade
sta	station
stab.	stable
stat	statistical
statist	statistician
std	standard
std	standard [Canadian]
stoch	stochastic
stoich	stoichiometric
s.t.p.	standard temperature and pressure [British]

STP	standard temperature and pressure	
str	straight	
str	strength	
sub	submarine	
sub	substitute	
sub.	substitution	
subl	sublimes	
subtr	subtraction	
Sun.	Sunday	
sup	supremum	
suppl	supplement	
susp	suspension	
s.v.	sub verbo (under the word)	
s.v.	sub voce (under the word)	
svp	saturated vapor pressure	
SVTP	sound velocity, temperature, and pressure	
sw	switch	
SW	short wave	
SW	southwest	
S-wave	shear wave	
Swed.	Swedish	
swr	standing-wave ratio	
sym	symbol	
sym	symmetrical	
symp	symposium	
syn.	synonym	
synth	synthesis	
t	time	
t	tonne (1000 kilograms)	
t	tonne (1000 kilograms) [British]	
t	tonne (1000 kilograms) [IUPAC]	
t	tonne (1000 kilograms) [IUPAP]	
t	troy	
T	tera ($=10^{12}$) (prefix) [International]	
T	tesla	
T	tesla [IUPAP]	
T	tritium	
Ta	tantalum	
tab.	tabulate	
tach	tachometer	
tan	tangent	
tan	tangent [British]	

tan	tangent [Canadian]	
tan	tangent [ISO]	
tan	tangent [IUPAP]	
tan^{-1}	inverse tangent	
tan^{-1}	inverse tangent [British]	
tanh	hyperbolic tangent	
tanh	hyperbolic tangent [British]	
tanh	hyperbolic tangent [Canadian]	
tanh	hyperbolic tangent [ISO]	
tanh	hyperbolic tangent [IUPAP]	
$tanh^{-1}$	inverse hyperbolic tangent	
$tanh^{-1}$	inverse hyperbolic tangent [British]	
Tb	terbium	
Tc	technetium	
Tc	teracycle	
Te	tellurium	
TE	thermal efficiency	
TE	transequatorial scatter	
TE	transverse electric	
technol	technological	
tel	telescope	
TEM	transverse electromagnetic	
temp	temperature	
temp	temperature [Canadian]	
temp.	temperature [British]	
TEM wave	transverse electromagnetic wave	
TENOC	Ten-Year Oceanographic Program	
tetr	tetragonal	
tetrah	tetrahedron	
tg	tangent [ISO]	
tg	tangent [IUPAP]	
TGA	thermogravimetric analysis	
TGA	thermogravimetric analyzer	
tgh	hyperbolic tangent [IUPAP]	
tgt	target	
th	theory	
Th	theorem	
Th	thorium	
T/H	tritium-hydrogen ratio	
therm	thermometer	
thermodyn	thermodynamics	
THI	temperature-humidity index	

Thurs.	Thursday	
Ti	titanium	
TID	traveling ionospheric disturbance	
Tl	thallium	
TL	transmission loss	
TLC	thin-layer chromatography	
TLE	thin-layer electrophoresis	
TLV	threshold limit value	
Tm	thulium	
TM	transverse magnetic	
TMA	theoretical mineral acidity	
TM wave	transverse magnetic wave	
TNP	terrestrial north pole	
tol	tolerance	
tonf	ton-force [British]	
tor	torsion	
TPP	terrestrial principal point	
tr	transition point	
tr.	transitive	
TR	technical report	
TRAAC	transit research and altitude control (satellite)	
traj	trajectory	
transl	translation	
transv	transversal	
transv	transverse	
trap.	trapezoid	
TRI	time-reversal invariant	
tricl	triclinic	
trig	trigonal	
trig	trigonometry	
TRIM	thin region integral method	
TRM	thermoremanent magnetization	
ts	tensile strength	
ts	tensile strength [Canadian]	
TSA	terrestrial spin-axis point	
TSP	terrestrial subpoint	
TT	thermomagnetic treatment	
Tues.	Tuesday	
tvp	true vapor pressure	
tw	true watt	
TWT	traveling-wave tube	
TYS	tensile yield strength	

u	atomic mass unit (unified)	
u	atomic mass unit (unified) [IUPAP]	
U	up	
U	uranium	
u.c.	unitary complex	
UDC	Universal Decimal Classification	
uf	ultrasonic frequency	
uhf	ultra-high frequency	
UK	United Kingdom	
univ	universe	
univ	university	
Univac	universal automatic computer	
unk	unknown	
UOV	units-of-variance	
u.s.	ubi supra (in the place above mentioned)	
USARP	United States Antarctic Research Program	
USSR	Union of Soviet Socialist Republics	
USW	ultra-short wave	
UT	universal time	
UTR	university training reactor	
u.v.	ultraviolet [British]	
uv	ultraviolet	
UVASER	ultraviolet amplification by stimulated emission of radiation	
UYP	upper yield point	
v	volt	
v	volt [Canadian]	
v.	verb	
v.	vide (see)	
V	vanadium	
V	volt [AIP]	
V	volt [British]	
V	volt [IUPAC]	
V	volt [IUPAP]	
va	volt-ampere [Canadian]	
VA	volt-ampere [British]	
VA	volt-ampere [LRL]	
vac	vacuum	
vac.	vacuum [British]	
val	value	
vap	vapor	

var	reactive volt-ampere	
var	reactive volt-ampere [Canadian]	
var	variable	
var	variance	
VC	volt-coulomb [British]	
VCM	vibrating-coil magnetometer	
v.d.	vapor density [British]	
vd	vapor density	
vel	velocity	
vers	versed sine	
vers	versed sine [Canadian]	
vgc	velocity gravity constant	
vhf	very-high frequency	
Vhn	Vickers hardness number	
V.I.	viscosity index	
visc	viscosity	
viz.	videlicet (namely)	
vlf	very-low frequency	
VM	voltmeter	
vol	volume	
vol.	volume [British]	
v.p.	vapor pressure [British]	
vp	vapor pressure	
vpm	vibrations per minute	
VP-meter	velocity-of-propagation meter	
Vpn	Vickers pyramid number	
vps	vibrations per second	
v.s.	vide supra (see above)	
vswr	voltage standing-wave ratio	
VT	vacuum tube	
w	watt	
w	watt [Canadian]	
W	tungsten	
W	watt [AIP]	
W	watt [British]	
W	watt [IUPAC]	
W	watt [IUPAP]	
W	west	
Wb	weber	
Wb	weber [British]	
Wb	weber [IUPAP]	
WB	wet bulb	

Wed.	Wednesday	
W-h	watt-hour [LRL]	
Wh	watt-hour	
Wh	watt-hour [British]	
whr	watt-hour [Canadian]	
WKF	well-known fact	
wl	wavelength	
WMS	World Magnetic Survey	
WNW	west-northwest	
w/o	weight percent	
WSW	west-southwest	
wt	weight	
wt	weight [Canadian]	
wt.	weight [British]	
Xe	xenon	
xform	transformation	
XH	extra-high	
xmsn	transmission	
xref	cross-reference	
xsect	cross-section	
xstr	transistor	
xu	x-unit	
Y	yttrium	
Yb	ytterbium	
yd	yard	
yd	yard [British]	
yd	yard [Canadian]	
yd	yard [ISO]	
yd^2	square yard	
yd^2	square yard [ISO]	
yd^3	cubic yard	
YP	yield point	
yr	year	
yr	year [Canadian]	
YS	yield strength	
ZD	zenith distance	
ZF	zero frequency	
ZGS	zero gradient synchrotron	
Zn	zinc	
Zr	zirconium	
ZT	zone time	
μ	micro ($=10^{-6}$) (prefix) [International]	

μ	micron
μ	micron [British]
μ	micron [Canadian]
μ	micron [ISO]
μ	micron [IUPAC]
μ²	square micron
μ²	square micron [Canadian]
μ³	cubic micron
μ³	cubic micron [Canadian]
μa	microampere [ASA]
μa	microampere [Canadian]
μA	microampere [AIP]
μA	microangstrom [NBS]
μA	microangstrom [AIP]
μbar	microbar [IUPAP]
μc	microcurie
μC	microcoulomb
μdyn	microdyne
μf	microfarad
μf	microfarad [Canadian]
μF	microfarad [AIP]
μg	microgram
μin.	microinch
μin.	microinch [Canadian]
μl	microliter
μliter	microliter [LRL]
μm	micrometer (measurement)
μm	micrometer (measurement) [British]
μm	micrometer (measurement) [ISO]
μM	micromole
μmin	microminute
μsec	microsecond
μv	microvolt
μv	microvolt [Canadian]
μV	microvolt [AIP]
μw	microwatt
μw	microwatt [Canadian]
μW	microwatt [LRL]
μμ	micromicro ($=10^{-12}$) (prefix, pico preferred)
μμ	micromicron
μμ	micromicron [Canadian]
μμf	micromicrofarad

μμf	micromicrofarad [Canadian]	
μμF	micromicrofarad [AIP]	
μμsec	micromicrosecond	
Ω	ohm	
Ω	ohm [British]	
Ω	ohm [Canadian]	
Ω	ohm [IUPAC]	
Ω	ohm [IUPAP]	

ABBREVIATIONS FOR USE IN TEXT
Alphabetically by Definition

abscissa	absc
absolute	abs
absolute [British]	abs.
absolute [Canadian]	abs
absolute ampere	abamp
absolute value of	abs
absorption	absorp
abstract	abstr
acceleration	accel
acceleration decibel	adb
acoustics	acous
actinium	Ac
activation energy	AE
actual mechanical advantage	AMA
acute	ac
addition	add.
ad finem (at the end)	ad fin.
ad infinitum (to infinity)	ad inf.
ad initium (at the beginning)	ad init.
ad interim (meanwhile)	ad int.
adjacent	adj
adjective	adj.
ad libitum (at pleasure)	ad lib.
ad locum (at the place)	ad loc.
Advanced Ballistic Re-entry Studies	ABRES
adverb	adv.
aerodynamic	aerodyn
agent	agt
aggregate	aggr
algebra	alg
algorithm	algo
alkali	alk
alkalinity	alky
alloy	aly
alternating current	ac
alternating current [British]	a.c.
alternating-current (adjective)	a-c
alternating-current (adjective only) [Canadian]	a-c

altitude	alt
aluminum	Al
American	Am.
American Standard	Amer Std
americium	Am
amorphous	amor
amount	amt
ampere	A
ampere [British]	A
ampere [British]	amp
ampere [Canadian]	amp
ampere [ISO]	A
ampere [IUPAC]	A
ampere [IUPAP]	A
ampere (combining form) [AAAS]	a
ampere-hour	Ah
ampere-hour [AIP]	A-h
ampere-hour [Canadian]	amp-hr
ampere-turn	At
ampere-turn [British]	AT
ampere-turn [LRL]	A-turn
ampere-turns per meter	At/m
ampere-turns per weber	At/Wb
amplitude, an elliptic function	am.
amplitude, an elliptic function [Canadian]	am
amplitude modulation	AM
analysis	anal
angle, side, angle	a.s.a.
angstrom	Å
angstrom [ASA]	Å
angstrom [British]	Å
angstrom [Canadian]	Å
angstrom [ISO]	Å
angstrom [IUPAC]	Å
angstrom [IUPAP]	Å
anno Domini (in the year of our Lord)	A.D.
ante meridiem (before noon)	a.m.
antilogarithm	antilog
antilogarithm [British]	antilog
antilogarithm [Canadian]	antilog
antimony	Sb
antitrigonometric	antitrig

antonym	ant.
apostilb	asb
appendix	appx
application	appl
Applied Mathematics Series [NBS]	AMS
approximate	approx
approximate [British]	approx.
approximation	approxn
April	Apr.
aqua bulliens (boiling water)	aq. bull.
aqua destillata (distilled water)	aq. dest.
argon	Ar
argument	arg
argument of	arg
argument of [ISO]	arg
argument of complex number [British]	arg
arithmetic	arith
arithmetic progression	AP
arsenic	As
article	art.
assistant	asst
astatine	At
astronomical	astron
astronomical unit	AU
astronomical unit [AAS]	a.u.
astronomic unit [ISO]	AU
astronomy	astron
astrophysics	astrophys
asymmetric	asym
asymptote	asymp
atmosphere	atm
atmosphere [British]	atm
atmosphere [Canadian]	atm
atmosphere [IUPAP]	atm
atmosphere, standard [AIP]	A_s
atmosphere, standard [British]	atm
atmospheric	atm
atmospheric [British]	atm.
atmospheric pressure	atm press.
atom connectivity matrix	ACM
atom connectivity matrix characteristic polynomial	ACMCP
atomic	at.

DPMA

atomic mass unit	amu
atomic mass unit (unified)	u
atomic mass unit (unified) [IUPAP]	u
atomic number	at. no.
atomic time	AT
atomic unit	au
atomic volume	at. vol
atomic weight	at. wt
atomic weight [British]	at. wt.
atomic weight [Canadian]	at. wt
atomic weight unit	awu
attenuation factor	AF
atto ($=10^{-18}$) (prefix) [International]	a
audio frequency	af
audio-frequency (adjective) [AIP]	af
August	Aug.
automatic	auto
automatic data processing	ADP
automatic recorder of static deformation	ARD
average	avg
average [AIP]	av
average [Canadian]	avg
average deviation	AD
avoirdupois	avdp
avoirdupois [Canadian]	avdp
axiom	Ax
axis	ax.
azimuth	az
azimuth [Canadian]	az
balance	bal
ballistic	bal
bandwidth	BW
bar	(spell)
bar [British]	b
bar [IUPAC]	b
Bardeen-Cooper-Schrieffer theory of superconductivity	BCS theory
barium	Ba
barn	b
barn [IUPAP]	b
battery	bat.
Baumé	Bé

DPMA

before Christ	B.C.
before present	B.P.
bel	(spell)
bel (combining form)	b
Belgian	Belg.
bellows	blws
berkelium	Bk
beryllium	Be
beryllium physics reactor	BPR
biaxial	biax
biconditional	bicond
billion electron volts	bev
billion electron volts [AIP]	BeV
billion electron volts [NBS]	Bev
binary	bin.
binary coded decimal	BCD
binary logarithm [ISO]	lb
binary logarithm [ISO]	$\log_2$
binary logarithm [IUPAP]	lb
binary logarithm [IUPAP]	$\log_2$
binomial	binom
binomial expansion	binom exp
binomial theorem	binom th
biochemistry	biochem
biot	Bi
bipyramidal	bipyr
birectangular	birect
bisector	bis
bismuth	Bi
black body	bb
body-centered (crystals)	b.c.
body-centered cubic (crystals)	b.c.c.
body-centered cubic (crystals) [AIP]	bcc
boiling nuclear superheat (reactor)	BONUS
boiling point	bp
boiling point [British]	b.p.
boiling point [Canadian]	bp
boiling-water reactor	BWR
boils at	b.
book	bk
boron	B
boundary	bdy

brake horsepower	bhp
brake horsepower [British]	b.h.p.
brake horsepower [Canadian]	bhp
breakdown voltage	BV
Brinell hardness number	Bhn
Brinell hardness number [Canadian]	Bhn
British	Brit.
British Standard	Br Std
British Standard [British]	B.S.
British thermal unit	Btu
British thermal unit [British]	Btu
British thermal unit [Canadian]	Btu
brittle fracture	BF
bromine	Br
built-up edge	BUE
bulletin	bull.
cadmium	Cd
calcium	Ca
calculate	calc
calculated	calc
calculated [British]	calc.
calculator	calc
calculus	calc
calibrate	cal
calibration	calbr
californium	Cf
calorie	cal
calorie [British]	cal
calorie [Canadian]	cal
calorie [IUPAC]	cal
calorie [IUPAP]	cal
Calorie (large)	Cal
Canadian	Can.
cancel	canc
candela	cd
candela [British]	cd
candela [ISO]	cd
candela [IUPAC]	cd
candela [IUPAP]	cd
candelas per square meter	cd/m^2
candle	c
candle [Canadian]	c

candlepower	cp
candlepower [Canadian]	cp
candles per square meter	c/m²
cantilever	cantil
capacitor	C
capacitor	cap.
capacity	cap.
capitals and small capitals	c.&s.c.
carbon	C
carbon copy	cc
cardinal	card.
card-programmed calculator	CPC
casting	cstg
catalog	cat.
catalyst	cat.
category	cat.
cathode ray	CR
caustic	caust
cavity	cav
celestial	cel
celestial north pole	CNP
celestial principal point	CPP
celestial spin-axis point	CSA
celestial subpoint	CSP
celestial telescope	celescope
Celsius	C
Celsius degree (temperature difference) [AAAS]	C°
center	ctr
center of buoyancy	cb
center of flotation	cf
center of gravity	cg
center of gravity [British]	c.g.
center of mass	c.m.
center of pressure	cp
centi ($=10^{-2}$) (prefix) [International]	c
centigrade	C
centigrade heat unit	chu
centiliter	cl
centiliter [Canadian]	cl
centimeter	cm
centimeter [British]	cm
centimeter [Canadian]	cm

centimeter [IUPAP]	**cm**
centimeter-gram-second	**cgs**
centimeter-gram-second [Canadian]	**cgs**
centimeter-gram-second electromagnetic	**cgsm**
centimeter-gram-second electrostatic	**cgse**
centimeter-gram-second unit	**cgsu**
centimeter of mercury	**cm Hg**
centimeter of oil	**cm oil**
centimeters per minute	**cm/min**
centimeters per second	**cm/sec**
centipoise	**cP**
centipoise [British]	**cP**
centipoise [NBS]	**cp**
centistoke	**cS**
centistoke [British]	**cS**
central limit theorem	**CLT**
centrifugal	**cent.**
centrifugal force	**c.f.**
century	**cent.**
cerium	**Ce**
cesium	**Cs**
channel	**chan**
Chapman function	**ch χ**
chapter	**chap.**
charge	**chg**
check	**chk**
chemical	**chem**
chemical [Canadian]	**chem**
chemically pure	**cp**
chemically pure [Canadian]	**cp**
chemist	**chem**
chemistry	**chem**
chlorine	**Cl**
chord	**chd**
chromium	**Cr**
circa (about)	**c.**
circa (about)	**ca.**
circle	**cir**
circuit	**ckt**
circular	**cir**
circular [Canadian]	**cir**
circular error	**CE**

circular error average	CEA
circular error probable	CEP
circular-mil foot	cir-mil ft
circular mils	cir mils
circular mils [Canadian]	cir mils
circulation	circ
circumference	**circ**
clarification	clar
clarify	clar
class	cl
classification	class.
clockwise	CW
close	cl
close-packed (crystals)	**c.p.**
close-packed hexagonal (crystals)	**c.p.h.**
coaxial	coax
coaxial thermal converter	CTC
cobalt	Co
cobalt bomb	C-bomb
coefficient	coef
coefficient [AIP]	coeff
coefficient [British]	coeff.
coefficient [Canadian]	coef
coherent	coh
cold-rolled	CR
collimator	collm
colloid	coll
colloidal	coll
cologarithm	colog
cologarithm [Canadian]	colog
Color Index	C. I.
columbium	Cb
column	col
combination	comb.
combine	comb.
combustion	comb.
Committee	Comm.
common	com
common logarithm [ISO]	lg
common logarithm [ISO]	log
common logarithm [ISO]	$\log_{10}$

common logarithm [British]	**log**
common logarithm [British]	**log₁₀**
communication	**comm**
commutator	**comm**
companion	**comp**
compass	**comp**
compensate	**compen**
compilation	**compil**
complement	**compl**
complementary	**compl**
complex conjugate	**c.c.**
complex loading	**CL**
component	**comp**
composite	**comp**
composition	**comp**
compound	**comp**
compound [British]	**cpd.**
computation	**comp**
concentrate	**conc**
concentrate [Canadian]	**conc**
concentrated	**concd**
concentrated [British]	**conc.**
concentration	**concn**
concentration [British]	**concn.**
concentric	**conc**
conclusion	**concl**
condenser	**cond**
condition	**cond**
conditional	**cond**
conductivity	**cond**
conductivity [Canadian]	**cond**
confer (compare)	**cf.**
conference	**conf**
configuration	**config**
congestion theory	**CGT**
congruent	**congr**
conjugate	**conj**
conjunction	**conj**
connect	**conn**
connection	**conn**
consecutive	**consec**
constant	**const**

constant [British]	const.
constant [Canadian]	const
construction	constr
contact potential difference	**cpd**
continue	cont
continuous wave	**CW**
contribution	contrib
control	cont
controlled thermonuclear fusion	**CTNF**
convention	conv
convergence	conv
convert	conv
cooling rate	**CR**
coordinate	coord
copper (cuprum)	**Cu**
coriolis force	**CF**
corollary	coroll
correction	corr
correlation	correl
correspond	corr
correspondence	corr
corresponding	corr
cosecant	csc
cosecant [British]	cosec
cosecant [Canadian]	csc
cosecant [ISO]	cosec
cosecant [IUPAP]	cosec
cosine	cos
cosine [British]	cos
cosine [Canadian]	cos
cosine [ISO]	cos
cosine [IUPAP]	cos
cosine integral function	**Ci**
cosine of the amplitude, an elliptic function	cn
cosine of the amplitude, an elliptic function [Canadian]	cn
cosmic noise absorption	**CNA**
COSPAR International Reference Atmosphere	**CIRA**
cotangent	cot
cotangent	ctn
cotangent [British]	cot
cotangent [Canadian]	cot

cotangent [ISO]	**cot**
cotangent [ISO]	**ctg**
cotangent [IUPAP]	**cot**
cotangent [IUPAP]	**ctg**
coulomb	**C**
coulomb [British]	**C**
coulomb [IUPAC]	**C**
coulomb [IUPAP]	**C**
coulombs per square meter	**C/m²**
counterclockwise	**CCW**
counter electromotive force	**cemf**
counter electromotive force [Canadian]	**cemf**
counts per minute	**counts/min**
counts per second	**counts/sec**
couple	**cpl**
coversed sine	**covers**
criterion	**crit**
critical	**crit**
critical [British]	**crit.**
critical degree of deformation	**CDD**
critical experiment	**CX**
cross-reference	**xref**
cross-section	**xsect**
cryogenics	**cryog**
crystal	**cryst**
cubic	**cu**
cubic [Canadian]	**cu**
cubic centimeter	**cc**
cubic centimeter	**cm³**
cubic centimeter [British]	**cm³**
cubic centimeter [Canadian]	**cm³**
cubic centimeter [Canadian]	**cu cm**
cubic centimeters per minute per kilogram	**cc/min/kg**
cubic decimeter	**dm³**
cubic foot	**ft³**
cubic foot [British]	**ft³**
cubic foot [Canadian]	**cu ft**
cubic foot [ISO]	**ft³**
cubic inch	**in.³**
cubic inch [British]	**in³**
cubic inch [Canadian]	**cu in.**
cubic inch [ISO]	**in³**

cubic meter	m³
cubic meter [British]	m³
cubic meter [Canadian]	cu m
cubic meter [Canadian]	m³
cubic meter [ISO]	m³
cubic micron	µ³
cubic micron [Canadian]	cu mu
cubic micron [Canadian]	cu µ
cubic micron [Canadian]	µ³
cubic millimeter	mm³
cubic millimeter [Canadian]	cu mm
cubic millimeter [Canadian]	mm³
cubic yard	yd³
cumulative	cum
curie	c
curie [AIP]	Ci
curie [British]	c
curium	Cm
current density	cd
current density [British]	c.d.
curve	crv
cycle (combining form)	c
cycles per second	cps
cycles per second [ASA]	c
cycles per second [British]	c/s
cycles per second [Canadian]	c
cycles per second (combining form)	c
cyclotron	cyclo
cylinder	cyl
cylinder [Canadian]	cyl
cylindrical	cyl
Czechoslovakian	Czech.
Danish	Dan.
dark	dk
dated	dtd
day [British]	d
day [ISO]	d
day [IUPAP]	d
debye	D
decadic logarithm [IUPAP]	lg
decadic logarithm [IUPAP]	log
December	Dec.

deci (=10^{-1}) (prefix) [International]	**d**
decibel	**db**
decibel [AIP]	**dB**
decibel [British]	**dB**
decibel [Canadian]	**db**
decibel [ISO]	**dB**
decigram	**dg**
deciliter	**dl**
decimal	**dec**
decimal classification	**DC**
decimal fraction	**DF**
decimeter	**dm**
declination	**dec**
decrease	**decr**
deduct	**ded**
definite	**def**
definition	**def**
deflect	**defl**
deflection	**defl**
degree	**deg**
degree [British]	**deg.**
degree [Canadian]	**deg**
degree Baumé [AIP]	**°B**
degree Celsius	**°C**
degree Celsius [IUPAC]	**°C**
degree Celsius (temperature interval) [British]	**degC**
degree Celsius (temperature value) [British]	**°C**
degree centigrade	**°C**
degree centigrade [ASA]	**C**
degree centigrade [Canadian]	**C**
degree Fahrenheit	**°F**
degree Fahrenheit [ASA]	**F**
degree Fahrenheit [Canadian]	**F**
degree Fahrenheit [IUPAC]	**°F**
degree Fahrenheit (temperature interval) [British]	**degF**
degree Fahrenheit (temperature value) [British]	**°F**
degree Kelvin	**°K**
degree Kelvin [ASA]	**K**
degree Kelvin [Canadian]	**K**
degree Kelvin [ISO]	**°K**
degree Kelvin [IUPAC]	**°K**
degree Kelvin [IUPAP]	**°K**

degree Kelvin (temperature interval) [British]	degK
degree Kelvin (temperature value) [British]	°K
degree of freedom	d.f.
degree of substitution	DS
deionizer	deion
deka (=10) (prefix)	dk
dekagram	dkg
dekaliter	dkl
delete	dele
deliquescent	deliq
delta amplitude, an elliptic function	dn
delta amplitude, an elliptic function [Canadian]	dn
demonstration	Dem
denominate	denom
denominator	denom
department	dept
deposit	dep
depression	depr
derivation	deriv
derivative	deriv
describe	descr
design	des
designation	desig
detecting magnetometer	DM
determinant	det
determinant of	det
determine	detm
deuterium	D
deuterium-deuterium reaction	D-D
deuterium-hydrogen ratio	D/H
deuterium-tritium reaction	D-T
deuteron	d
deuteron-deuteron (reactor)	DD
deuteron-triton (reactor)	DT
develop	dev
deviate	dev
deviation	dev
dew point	dp
diagonal	diag
diagram	diag
diameter	diam
diameter [ASA]	dia

diameter [Canadian]	diam
diaphragm	diaph
dictionary	dict
dielectric loading factor	DLF
difference	diff
differential	diff
differential thermoanalyzer	DTA
diffraction	diffr
digital computer	dig comp
digital data	DD
digital-to-analog	D/A
dilute	dil
dilute [British]	dil.
dilute solution viscosity	DSV
dimension	dim.
dimorphous	dimorph
diopter	diop
direct	dir
direct current	dc
direct-current (adjective)	d-c
direct current (adjective only) [Canadian]	d-c
direct current [British]	d.c.
direct-current experimental device	DCX
direction	dir
director	dir
discharge	disch
discriminant	discrim
discrimination	disc.
disintegrations per second	dps
disintegrations per second [AIP]	dis/sec
disjunction	disj
disodium phosphate	DSP
displacement	displ
distance	dist
distortion	distn
distribution	distr
distribution factor (radiation)	DF
divergence	div
divergence of	div
divide	div
division	div
document	doc

down	D
dram	**dr**
dram [Canadian]	dr
drawing	dwg
drive	**dr**
drum	dr
dry bulb	DB
dynamic	dyn
dyne	dyn
dyne [AAAS]	dy
dyne [British]	dyn
dyne [IUPAC]	dyn
dyne [IUPAP]	dyn
dysprosium	Dy
east	E
east-northeast	ENE
east-southeast	ESE
east-west	E-W
eccentric	**ecc**
eccentric orbiting geophysical observatory	EGO
echelon	ech
economist	econ
edition	ed
editor	ed.
efficiency	**eff**
efficiency [Canadian]	eff
einsteinium	E
elasticity	elast
elastic limit	EL
electric	elec
electric [Canadian]	elec
electrical engineer	EE
electric horsepower	ehp
electrochemistry	electrochem
electroluminescence	EL
electroluminescent-photoconductive	EL-PC
electroluminescent-photoresponsive	EL-PR
electrolyte	elect.
electromagnetic	EM
electromagnetic amplifying lens	EAL
electromagnetic radiation	EMR
electromagnetic unit	emu

electromagnetic unit [British]	**e.m.u.**
electromotive force	**emf**
electromotive force [British]	**e.m.f.**
electromotive force [Canadian]	**emf**
electron capture	**EC**
electronic data processing	**EDP**
electronic data processing equipment	**EDPE**
electronic digital computer	**EDC**
electronics	**elect.**
electron microscope	**EM**
electron paramagnetic resonance	**EPR**
electron spin resonance	**ESR**
electron volt	**ev**
electron volt [AIP]	**eV**
electron volt [British]	**eV**
electron volt [IUPAP]	**eV**
electrophysics	**electrophys**
electrostatic	**es**
electrostatic unit	**esu**
electrostatic unit [British]	**e.s.u.**
element	**elem**
elementary transformation matrix	**exm**
elevation	**el**
elevation [Canadian]	**el**
eliminate	**elim**
ellipse	**ell.**
elongation	**elong**
emission	**emiss**
emulsion	**emul**
engine	**eng**
engineer	**engr**
entropy unit	**eu**
enumeration	**enum**
envelope	**env**
ephemeris time	**ET**
equation	**eq**
equation [AIP]	**Eq.**
equation [British]	**eqn.**
equation [Canadian]	**eq**
equations [AIP]	**Eqs.**
equatorial ring current	**ERC**
equilibrium	**equil**

equipment	equip.
equivalent	equiv
equivalent [British]	equiv.
equivalent direct radiation	EDR
erbium	Er
erg	(spell)
error	E
error function	erf
error function, complementary	erfc
error function of [British]	erf
establishment	estab
estimate	est
et alibi (and elsewhere)	et al.
et alii (and others)	et al.
et cetera	etc.
ether	eth.
et sequens (and the following)	et seq.
europium	Eu
evaporate	evap
evolution	evol
evolves	ev
examination	exam
example	ex
exchange	exch
excitation	exc
exempli gratia (for example)	e.g.
expand	exp
expansion	exp
experiment	expt
experiment [British]	expt.
experimental	exptl
experimental [British]	expt.
experimental beryllium oxide reactor	EBOR
experimental boiling-water reactor	EBWR
experimental breeder reactor	EBR
experimental gas-cooled reactor	EGCR
experimental organic-cooled reactor	EOCR
explanation	expl
exponential	exp
exponential function of	exp
exponential function of [British]	exp
exponential integral function	Ei

exposure ... **expos**
exposure factor ... **EF**
expression ... **expr**
expulsion ... **exp**
exsecant ... **exsec**
extension ... **ext**
extent ... **ext**
exterior ... **ext**
extra-high ... **XH**
extreme ... **extr**
extremely-high frequency ... **ehf**
Fabry-Perot spherical (interferometer) ... **FPS**
face-centered (crystals) ... **f.c.**
face-centered cubic (crystals) ... **f.c.c.**
face-centered cubic (crystals) [AIP] ... **fcc**
Fahrenheit ... **F**
Fahrenheit degree (temperature difference) [AAAS] ... **F°**
failed ... **F**
fallacy ... **fall.**
false ... **F**
farad ... **f**
farad [AAAS] ... **fd**
farad [AIP] ... **F**
farad [British] ... **F**
farad [Canadian] ... **f**
farad [IUPAC] ... **F**
farad [IUPAP] ... **F**
fast-breeder reactor ... **FBR**
February ... **Feb.**
feet per second ... **fps**
feet per second [Canadian] ... **fps**
feet per second [ISO] ... **ft/s**
feet per second squared [ISO] ... **ft/s^2**
femto ($=10^{-15}$) (prefix) [International] ... **f**
fermi ... **F**
fermium ... **Fm**
fiber ... **fbr**
Figure ... **Fig.**
figure [British] ... **fig.**
filament ... **fil**
fine structure ... **fs**
Finnish ... **Finn.**

first sine-integral function	Si
fission product [AEC]	F.P.
float	flt
floating index point	FLIP
floating laboratory instrument platform	FLIP
floor	fl
fluorine	F
fluid	fl
fluid [Canadian]	fl
fluid dram	fl dr
fluid ounce	fl oz
fluorescent	fluor
foot [British]	ft
foot [Canadian]	ft
foot [ISO]	ft
foot-candle	ft-c
foot-candle [Canadian]	ft-c
foot-lambert	ft-L
foot-lambert [AAAS]	ft-lam
foot-lambert [Canadian]	ft-L
foot-pound	ft-lb
foot-pound [Canadian]	ft-lb
foot-pound-force	ft-lbf
foot-pound-force [British]	ft lbf
foot-pound-second	fps
foot-pound-second [Canadian]	fps
foot-pound-second unit	fpsu
force	F
force-length-time system	FLT
force-mass-length-time system	FMLT
formula	form.
forward	fwd
forward propagation by ionospheric scatter	FPIS
forward propagation by tropospheric scatter	FPTS
four-dimensional	4-D
Fourier series	FS
fraction	frac
fractional	frac
fractional horsepower	fhp
francium	Fr
franklin	Fr
freezing point	fp

freezing point [British]	**f.p.**
freezing point [Canadian]	**fp**
French	**Fr.**
frequency	**freq**
frequency modulation	**FM**
fresnel	**fr**
Friday	**Fri.**
frontier set of	**Fr**
fulcrum	**ful**
full period	**FP**
full width at half maximum	**FWHM**
fusion point	**fnp**
fusion point [Canadian]	**fnp**
gadolinium	**Gd**
gage	**ga**
gal	**Gal**
gal [ISO]	**Gal**
gallium	**Ga**
gallon	**gal**
gallon [British]	**gal**
gallon [Canadian]	**gal**
Galois field	**GF**
galvanometer	**galv**
Gamow-Teller (selection rules for beta decay)	**G-T**
gas-cooled reactor	**GCR**
gauss	**G**
gauss [British]	**G**
gauss [IUPAP]	**G**
Geiger-Muller (radiation counter)	**GM**
general	**gen**
generator	**gen**
geochemistry	**geochem**
geodetic	**geod**
geological	**geol**
geomagnetic electrokinetograph	**GEK**
geomagnetism	**geomag**
geometric	**geom**
geometric mean	**GM**
geometric progression	**GP**
geometry	**geom**
geophysics	**geophys**
geopotential meter	**gpm**

German	Ger.
germanium	Ge
giga (=10⁹) (prefix) [International]	G
gigacycle	Gc
gigaelectron volt	Gev
gigaelectron volt [LRL]	GeV
gigahertz	GHz
gilbert	Gi
globular	glob
glossary	gloss.
gold (aurum)	Au
grade	gr
gradient	grad
gradient of	grad
grain	gr
grain [British]	gr
gram	g
gram [British]	g
gram [Canadian]	g
gram [IUPAC]	g
gram [IUPAP]	g
gram-force	gf
gram-mass	gm
gram-meter	g-m
gram-molecular volume	gmv
gram-molecule [British]	mole
grams per cubic centimeter	g/cc
grams per cubic centimeter	g/cm³
granular	gran
grating	grtg
gravity	g
greatest common divisor	gcd
greatest common divisor	G.C.D.
greatest common divisor [Canadian]	gcd
greatest common factor	gcf
greatest common factor	G.C.F.
greatest common subgroup	gcs
greatest common subgroup	G.C.S.
greatest lower bound	g.l.b.
Greenwich apparent time	GAT
Greenwich hour angle	GHA
Greenwich mean sidereal time	GMST

Greenwich mean time	**GMT**
Greenwich sidereal time	**GST**
grid	**g**
group	**gp**
guaranteed yield strength	**GYS**
Gudermannian	**gd**
Guide Line Identification Program for Antimissile Research	**GLIPAR**
hafnium	**Hf**
handbook	**hdbk**
harmonic mean	**HM**
harmonic progression	**HP**
haversine	**hav**
haversine [Canadian]	**hav**
heat	**ht**
heat input	**H.I.**
heavy ion linear accelerator	**HILAC**
heavy water components test reactor	**HWCTR**
heavy-water-moderated gas-cooled reactor	**HWGCR**
hecto ($=10^2$) (prefix)	**h**
hectogram	**hg**
hectoliter	**hl**
helium	**He**
henry [AIP]	**H**
henry [British]	**H**
henry [Canadian]	**h**
henry [IUPAC]	**H**
henry [IUPAP]	**H**
hertz	**Hz**
hertz [AAAS]	**hz**
hertz [ISO]	**Hz**
hertz [IUPAC]	**Hz**
hertz [IUPAP]	**Hz**
heterogeneous	**heterog**
hexagon	**hex**
hexagonal close-packed (crystals)	**h.c.p.**
hexagonal close-packed (crystals) [AIP]	**hcp**
high current density	**hcd**
highest common factor	**hcf**
highest common factor	**H.C.F.**
high frequency	**hf**
high pressure	**hp**
high-temperature gas-cooled reactor	**HTGCR**

high-temperature gas reactor	HTGR
high tensile strength	HTS
holmium	Ho
homogeneous	homo
hone	hn
honed	hnd
honorary	hon.
horizon	hor
horsepower	hp
horsepower [British]	hp
horsepower [Canadian]	hp
hot	h.
hot-rolled	HR
hot-water reactor	HWR
hour	h
hour [British]	h
hour [Canadian]	hr
hour [ISO]	h
hour [IUPAC]	h
hour [IUPAP]	h
hour (in astronomical tables)	h
hour (in astronomical tables) [Canadian]	h
hour angle	HA
hour angle declination	HADEC
hundred	C
hundred [Canadian]	C
hydraulic	hyd
hydrodynamic	hydrodyn
hydroelectric	hydroelec
hydrogen	H
hydrogen-ion concentration	pH
hydrogen-ion concentration [British]	pH
hydrology	hydrol
hydrolyzed	hyd
hydromechanical	hydromech
hygroscopic	hyg
hyperbola	hyperb
hyperbolic	hyperbol
hyperbolic cosecant	csch
hyperbolic cosecant [British]	cosech
hyperbolic cosecant [ISO]	cosech
hyperbolic cosecant [IUPAP]	cosech

DPMA

hyperbolic cosine	cosh	
hyperbolic cosine [British]	cosh	
hyperbolic cosine [Canadian]	cosh	
hyperbolic cosine [ISO]	**cosh**	
hyperbolic cosine [IUPAP]	cosh	
hyperbolic cotangent	coth	
hyperbolic cotangent	**ctnh**	
hyperbolic cotangent [British]	coth	
hyperbolic cotangent [ISO]	coth	
hyperbolic cotangent [IUPAP]	coth	
hyperbolic cotangent [IUPAP]	ctgh	
hyperbolic secant	sech	
hyperbolic secant [British]	sech	
hyperbolic secant [ISO]	sech	
hyperbolic secant [IUPAP]	sech	
hyperbolic sine	sinh	
hyperbolic sine [British]	sinh	
hyperbolic sine [Canadian]	sinh	
hyperbolic sine [ISO]	sinh	
hyperbolic sine [IUPAP]	sinh	
hyperbolic tangent	tanh	
hyperbolic tangent [British]	tanh	
hyperbolic tangent [Canadian]	tanh	
hyperbolic tangent [ISO]	tanh	
hyperbolic tangent [IUPAP]	tanh	
hyperbolic tangent [IUPAP]	tgh	
hyperfine structure	hfs	
hypersonic test vehicle	HTV	
hypotenuse	hyp	
hypothesis	hyp	
ibidem (in the same place)	ibid.	
ideal mechanical advantage	IMA	
idem (the same)	id.	
idem quod (the same as)	i.q.	
identical	ident	
id est (that is)	i.e.	
illustration	illus	
image principal point	IPP	
image spin-axis point	ISA	
imaginary	imag	
imaginary part of	Im	
imaginary part of [British]	Im	

imaginary part of [ISO]	Im
imaginary part of [IUPAP]	Im
impact	imp.
impact energy	IE
impact strength	IS
impulse	imp.
inch	in.
inch [British]	in
inch [Canadian]	in.
inch [ISO]	in
inches per second	ips
inches per second [Canadian]	ips
inch of mercury	in. Hg
inch of mercury [British]	inHg
inch of oil	in. oil
inch-pound [ASA]	in-lb
inch-pound [Canadian]	in-lb
inch-pound [LRL]	in.-lb
inclination	incl
including	incl
inclusive	incl
incoherent	incoh
increase	incr
increment	incr
indefinite	indef
independent	indep
indeterminate	indet
index	inx
index error	IE
indium	In
individual	indiv
inductance	ind
inductance-capacitance	LC
inductance-capacitance-resistance	LCR
induction	ind
industry	ind
inequality	ineq
inference	infer.
infinity	inf
infinum	inf.
inflammable	inflam
inflection	infl

information	info
infrared	ir
infrared [British]	i.r.
infrared amplification by stimulated emission of radiation	iraser
infrared measurement program	IRMP
initial	init
initial vapor pressure	ivp
inorganic	inorg
inscribed angle	I.A.
inscribed circle	I.C.
inside diameter	i.d.
inside radius	i.r.
insoluble	insol
institute	inst
instructor	instr
integer	int
integral	int
integrate	int
integrated neutron flux	nvt
Inter-American Geodetic Survey	IAGS
intercept	int
interest	int
interfacial tension	IFT
interior	int
intermediate frequency	i.f.
internal	int
internal [Canadian]	int
internal friction	IF
international	intl
international angstrom	IA
International Astrophysical Decade	IAD
international brightness coefficient	IBC
International Cooperative Emulsion Flight (study)	ICEF
International Critical Tables	ICT
International Geophysical Cooperation	IGC
International Geophysical Year	IGY
International Map of the World on the Millionth Scale	IMW
international nautical mile	i.n. mi.
International Polar Year	IPY

International Satellites for Ionospheric Studies (program)	ISIS
international unit	IU
International World Day Service	IWDS
International Year of the Quiet Sun	IQSY
intersect	int
interval	int
invariant	invar
inverse	inv
inverse cosecant	arc csc
inverse cosecant	csc^{-1}
inverse cosecant [British]	arc cosec
inverse cosecant [British]	$cosec^{-1}$
inverse cosecant [ISO]	arccosec
inverse cosecant [IUPAP]	arccosec
inverse cosine	arc cos
inverse cosine	cos^{-1}
inverse cosine [British]	arc cos
inverse cosine [British]	cos^{-1}
inverse cosine [ISO]	arccos
inverse cosine [IUPAP]	arccos
inverse cotangent	arc cot
inverse cotangent	arc ctn
inverse cotangent	cot^{-1}
inverse cotangent	ctn^{-1}
inverse cotangent [British]	arc cot
inverse cotangent [British]	cot^{-1}
inverse cotangent [ISO]	arccot
inverse cotangent [ISO]	arcctg
inverse cotangent [IUPAP]	arccot
inverse cotangent [IUPAP]	arcctg
inverse hyperbolic cosecant	$csch^{-1}$
inverse hyperbolic cosecant [British]	arc cosec
inverse hyperbolic cosecant [British]	$cosech^{-1}$
inverse hyperbolic cosecant [ISO]	arcosech
inverse hyperbolic cosecant [IUPAP]	arcosech
inverse hyperbolic cosine	$cosh^{-1}$
inverse hyperbolic cosine [British]	arc cosh
inverse hyperbolic cosine [British]	$cosh^{-1}$
inverse hyperbolic cosine [ISO]	arcosh
inverse hyperbolic cosine [IUPAP]	arcosh

inverse hyperbolic cotangent	$\coth^{-1}$
inverse hyperbolic cotangent	$\ctnh^{-1}$
inverse hyperbolic cotangent [British]	arc coth
inverse hyperbolic cotangent [British]	$\coth^{-1}$
inverse hyperbolic cotangent [ISO]	arcoth
inverse hyperbolic cotangent [IUPAP]	arcoth
inverse hyperbolic cotangent [IUPAP]	arctgh
inverse hyperbolic secant	$\sech^{-1}$
inverse hyperbolic secant [British]	arc sech
inverse hyperbolic secant [British]	$\sech^{-1}$
inverse hyperbolic secant [ISO]	arsech
inverse hyperbolic secant [IUPAP]	arsech
inverse hyperbolic sine	$\sinh^{-1}$
inverse hyperbolic sine [British]	arc sinh
inverse hyperbolic sine [British]	$\sinh^{-1}$
inverse hyperbolic sine [ISO]	arsinh
inverse hyperbolic sine [IUPAP]	arsinh
inverse hyperbolic tangent	$\tanh^{-1}$
inverse hyperbolic tangent [British]	arc tanh
inverse hyperbolic tangent [British]	$\tanh^{-1}$
inverse hyperbolic tangent [ISO]	artanh
inverse hyperbolic tangent [IUPAP]	artanh
inverse hyperbolic tangent [IUPAP]	artgh
inverse secant	arc sec
inverse secant	$\sec^{-1}$
inverse secant [British]	arc sec
inverse secant [British]	$\sec^{-1}$
inverse secant [ISO]	arcsec
inverse secant [IUPAP]	arcsec
inverse sine	arc sin
inverse sine	$\sin^{-1}$
inverse sine [British]	arc sin
inverse sine [British]	$\sin^{-1}$
inverse sine [ISO]	arcsin
inverse sine [IUPAP]	arcsin
inverse tangent	arc tan
inverse tangent	$\tan^{-1}$
inverse tangent [British]	arc tan
inverse tangent [ISO]	arctan
inverse tangent [British]	$\tan^{-1}$
inverse tangent [ISO]	arctg
inverse tangent [IUPAP]	arctan

inverse tangent [IUPAP]	arctg
involute	invol
iodine	I
ion chamber	IC
ion rocket engine	IRE
iridium	Ir
iron (ferrum)	Fe
irradiation	irrad
isolate	isol
isomeric transition	IT
isometric	isom
isosceles	isos
isothermal	isoth
isotope dilution	ID
Italian	Ital.
January	Jan.
Japanese	Japan.
joule	J
joule [AIP]	J
joule [British]	J
joule [Canadian]	j
joule [IUPAC]	J
joule [IUPAP]	J
kaiser	K
Kansas	Kans.
Kelvin	K
Kelvin degree (temperature difference) [AAAS]	K°
key word in context (indexing system)	KWIC
kilo (=10^3) (prefix) [International]	k
kilobar	kb
kilocalorie	kcal
kilocalorie [British]	kcal
kilocalorie [Canadian]	kcal
kilocalorie [IUPAP]	kcal
kilocalories per mole	kcal/mole
kilocurie	kc
kilocurie [LRL]	kC
kilocycle	kc
kilocycles per second [AIP]	kc/sec
kilocycles per second [Canadian]	kc
kiloelectron volt	kev
kiloelectron volt [AIP]	keV

DPMA

kilogauss	kG
kilogram	kg
kilogram [British]	kg
kilogram [Canadian]	kg
kilogram [ISO]	kg
kilogram [IUPAP]	kg
kilogram-force	kgf
kilogram-force [British]	kgf
kilogram-mass	kgm
kilogram-meter	kg-m
kilogram-meter [Canadian]	kg-m
kilogram-mole	kg-mole
kilograms per cubic meter	kg/m^3
kilograms per cubic meter [Canadian]	kg/m^3
kilograms per cubic meter [Canadian]	kg per cu m
kilograms per second	kg/sec
kilograms per second [Canadian]	kgps
kilograms per square centimeter	kg/cm^2
kilogram-weight	kg-wt
kilohertz	kHz
kilohm	kΩ
kilojoule	kj
kilojoule [AIP]	kJ
kiloliter	kl
kiloliter [Canadian]	kl
kiloliter [LRL]	kliter
kilomega (=10^9) (prefix, giga preferred)	kM
kilometer	km
kilometer [Canadian]	km
kilo-oersted	kOe
kiloparsec	kpc
kilopounds	kips
kilovar; reactive kilovolt-ampere	kvar
kilovolt	kv
kilovolt [AIP]	kV
kilovolt [Canadian]	kv
kilovolt-ampere [AIP]	kVA
kilovolt-ampere [Canadian]	kva
kilowatt	kw
kilowatt [AIP]	kW
kilowatt [British]	kW
kilowatt [Canadian]	kw

kilowatt-hour [AIP]	kWh
kilowatt-hour [British]	kWh
kilowatt-hour [Canadian]	kwhr
kilowatt-hour [IUPAP]	kWh
kinetic	kin.
kinetic energy	KE
kinetic energy released in material	kerma
krypton	Kr
laboratory	lab
labyrinth	laby
lambert	L
lambert [AAAS]	lam
lambert [Canadian]	L
langley	ly
lanthanum	La
latent heat	lat ht
latent heat [British]	lat.ht.
lateral	lat
latitude	lat
latitude [Canadian]	lat
lawrencium	Lw
lead	Pb
leakage	lkg
least common factor	lcf
least common factor	L.C.F.
least common multiple	lcm
least common multiple	L.C.M.
least common multiple [Canadian]	lcm
lever	lvr
light	lt
light amplification by stimulated emission of radiation	laser
light-year	lt-yr
limit	lim
limit-in-mean	l.i.m.
limit of [British]	lim
limit of [ISO]	lim
line	l
linear	lin
linear accelerator	LINAC
linear energy transfer	LET
linear equation	lin eq

linear programming	LP
linear straining rate	LSR
line of sight	LOS
lines	ll
liquid	liq
liquid [British]	liq.
liquid [Canadian]	liq
liquid air accumulator rocket	LAAR
liter	l
liter [British]	l
liter [Canadian]	l
liter [ISO]	l
liter [IUPAC]	l
liter [IUPAP]	l
literature	lit.
lithium	Li
local apparent time	LAT
local hour angle	LHA
local mean time	LMT
local sidereal time	LST
local thermodynamics equilibrium	L.T.E.
local zone time	LZT
locate	loc
loco citato (in the place cited)	loc. cit.
locus	loc
logarithm	log
logarithmic integral function	li
logarithmic mean temperature difference	LMTD
logarithm to base 10	$\log_{10}$
logarithm to the base a [ISO]	$\log_a$
logarithm to the base a [IUPAP]	$\log_a$
logic	log.
longitude	long.
longitude [Canadian]	long.
longitudinal center of buoyancy	lcb
longitudinal center of flotation	lcf
longitudinal center of gravity	lcg
long range	LRG
love wave	Q-wave
lower case	l.c.
lower yield point	LYP
lowest common denominator	lcd

lowest common denominator	L.C.D.
lowest common multiple	lcm
lowest common multiple	L.C.M.
low frequency	lf
low pressure	lp
lumen	lm
lumen [British]	lm
lumen [Canadian]	l
lumen [IUPAC]	lm
lumen [IUPAP]	lm
lumen-hour	lm-hr
lumen-hour [Canadian]	l-hr
lumen-second	lm-sec
lumens per square meter	lm/m^2
lumens per watt [AIP]	lm/W
lumens per watt [ASA]	lpw
lumens per watt [Canadian]	lpw
lumens per watt [IEC]	lpW
lutetium	Lu
lux	lx
lux [British]	lx
lux [IUPAC]	lx
lux [IUPAP]	lx
machine	mach
magnesium	Mg
magnet	mag
magnetic moment	MM
magnetic north	MN
magnetic south	MS
magnetogasdynamics	MGD
magnetohydrodynamics	MHD
magnetomotive force	mmf
magnetomotive force [British]	m.m.f.
magnitude	mag
major	maj
major axis	maj ax.
manganese	Mn
manometer	manom
manual	man.
March	Mar.
margin of elastic energy	MEE
mark	mk

mass-length-time system	**MLT**
mathematical	**math**
mathematics	**math**
matrix	**mat.**
maximum	**max**
maximum [British]	**max.**
maximum [Canadian]	**max**
maxwell	**Mx**
maxwell [IUPAP]	**Mx**
mean deviation	**MD**
mean free path	**mfp**
mean horizontal candle	**mhc**
mean horizontal candlepower	**mhcp**
mean horizontal candlepower [Canadian]	**mhcp**
mean sea level	**MSL**
mean spherical candle	**msc**
mean spherical candlepower	**mscp**
mean square	**MS**
mean square difference	**MSD**
mean square error	**MSE**
mean variation	**MV**
mean zonal candlepower	**mzcp**
measure	**meas**
mechanical	**mech**
mechanical engineer	**ME**
mechanics	**mechs**
median	**med**
medium	**med**
medium frequency	**mf**
medium range	**MR**
mega (=10⁶) (prefix) [International]	**M**
megacurie	**Mc**
megacurie [LRL]	**MC**
megacycle	**Mc**
megacycles per second [AIP]	**Mc/sec**
megamega (=10¹²) (prefix, tera preferred)	**MM**
megaton	**MT**
megaton [LRL]	**Mt**
megawatt	**Mw**
megawatt [LRL]	**MW**
megohm	**MΩ**
megohm [AAAS]	**Mohm**

melt flow index	**MFI**
melting point	**mp**
melting point [British]	**m.p.**
melting point [Canadian]	**mp**
mendelevium	**Mv**
mercury	**Hg**
metabolism	**metab**
metallurgical	**met.**
Meteorological Rocket Network	**MRN**
meter	**m**
meter [British]	**m**
meter [Canadian]	**m**
meter [ISO]	**m**
meter [IUPAC]	**m**
meter [IUPAP]	**m**
meter-kilogram	**m-kg**
meter-kilogram [Canadian]	**m-kg**
meter-kilogram-second	**mks**
meter-kilogram-second-ampere	**mksa**
meter-kilogram-second electromagnetic unit	**mksm**
meter-kilogram-second unit	**mksu**
meters per second	**m/sec**
meters per second [British, I.E.E.]	**m/s**
meters per second [ISO]	**m/s**
meters per second squared [ISO]	**m/s^2**
method	**mthd**
method of rapid determination	**MRD**
micro ($=10^{-6}$) (prefix) [International]	**μ**
microampere [AIP]	**μA**
microampere [ASA]	**mu a**
microampere [ASA]	**μa**
microampere [Canadian]	**mu a**
microampere [Canadian]	**μa**
microangstrom [AIP]	**μÅ**
microangstrom [NBS]	**μÅ**
microbar [IUPAP]	**μbar**
microcoulomb	**μC**
microcurie	**μc**
microdyne	**μdyn**
microfarad	**μf**
microfarad [AIP]	**μF**
microfarad [Canadian]	**μf**

DPMA

microgram	μg
microinch	μin.
microinch [Canadian]	μin.
microliter	μl
microliter [LRL]	μliter
micrometer	mic
micrometer (measurement)	μm
micrometer (measurement) [British]	μm
micrometer (measurement) [ISO]	μm
micromicro ($=10^{-12}$) (prefix, pico preferred)	$\mu\mu$
micromicrofarad	$\mu\mu$f
micromicrofarad [AIP]	$\mu\mu$F
micromicrofarad [Canadian]	$\mu\mu$f
micromicron	$\mu\mu$
micromicron [Canadian]	mu mu
micromicron [Canadian]	$\mu\mu$
micromicrosecond	$\mu\mu$sec
microminute	μmin
micromole	μM
micron	μ
micron [British]	μ
micron [Canadian]	mu
micron [Canadian]	μ
micron [ISO]	μ
micron [IUPAC]	μ
microscope	micr
microsecond	μsec
microvolt	μv
microvolt [AIP]	μV
microvolt [Canadian]	μv
microwatt	μw
microwatt [Canadian]	mu w
microwatt [Canadian]	μw
microwatt [LRL]	μW
microwave	MW
microwave amplification by stimulated emission of radiation	maser
midmean	Mm
midpoint	midpt
mil	(spell)
miles per hour	mph
miles per hour [Canadian]	mph

DPMA

miles per hour [ISO]	**mile/h**
military	**mil**
milli ($=10^{-3}$) (prefix) [International]	**m**
milliampere	**ma**
milliampere [AIP]	**mA**
milliampere [Canadian]	**ma**
milliangstrom	**mÅ**
millibar	**mb**
millibar [British]	**mb**
millibarn	**mb**
millicurie	**mc**
millicurie [AIP]	**mCi**
millicurie [LRL]	**mC**
milliequivalent [ACS]	**meq.**
milligal	**mGal**
milligram	**mg**
milligram [Canadian]	**mg**
millihenry [AIP]	**mH**
millihenry [Canadian]	**mh**
millilambert	**mL**
millilambert [Canadian]	**mL**
milliliter	**ml**
milliliter [British]	**ml**
milliliter [Canadian]	**ml**
milliliter [LRL]	**mliter**
milli-mass-units	**mmu**
milli-mass-units [LRL]	**mMU**
millimeter	**mm**
millimeter [British]	**mm**
millimeter [Canadian]	**mm**
millimeter of mercury	**mm Hg**
millimeter of mercury [British]	**mmHg**
millimicro ($=10^{-9}$) (prefix, nano preferred)	**mμ**
millimicron	**mμ**
millimicron [Canadian]	**m mu**
millimicron [Canadian]	**mμ**
millimicrosecond	**mμsec**
millimole	**mM**
million electron volts	**Mev**
million electron volts [AIP]	**MeV**
million volts [AIP]	**MV**
millirem	**mrem**

milliroentgen	**mr**
milliroentgens per hour	**mr/hr**
milliroentgens per year	**mr/yr**
millisecond	**msec**
millivolt	**mv**
millivolt [AIP]	**mV**
millivolt [Canadian]	**mv**
milliwatt	**mw**
milliwatt [AIP]	**mW**
mineral	**min**
mineralogy	**mineral.**
minimum	**min**
minimum [British]	**min.**
minimum [Canadian]	**min**
minimum perceptible color difference	**MPCD**
minimum pure radium equivalent	**MPRE**
minor axis	**min ax.**
minute	**min**
minute [Canadian]	**min**
minute [ISO]	**min**
minute [IUPAC]	**min**
minute [IUPAP]	**min**
minute (time) [British]	**min**
minute (time, in astronomical tables)	**m**
minute (time, in astronomical tables) [Canadian]	**m**
miscellaneous	**misc**
miscible	**misc**
missile	**msl**
mixed fission products	**MFP**
mixture	**mix.**
mode	**Mo**
model	**mod**
moderate	**mod**
modulo	**mod**
modulus	**mod**
molar	**M**
molar (concentration) [British]	**M**
mole [IUPAP]	**mol**
mole [LRL]	**M**
molecular	**mol**
molecular [British]	**mol.**
molecular weight	**mol wt**

molecular weight [ASA]	mol. wt
molecular weight [British]	mol. wt.
molecular weight [Canadian]	mol. wt
molecule	mol
molecule [British]	mol.
mole percent	M%
molybdenum	Mo
Monday	Mon.
monograph	monog
month	mo
Moon-Earth plane	MEP
motor	mot
motor generator [LRL]	m.g.
mount	mt
mounting	mtg
multigroup internuclear slab transport	MIST
multiple	mult
multiplication	mult
mu-meson	muon
myria (=10^4) (prefix)	my
myriagram	myg
myrialiter	myl
nadir	N
nano (=10^{-9}) (prefix) [International]	n
nanocurie	nc
nanometer	nm
nanosecond	nsec
nanosecond [IUPAP]	ns
National Nuclear Energy Series (of AEC)	NNES
National Operational Meterological Satellite System	NOMSS
National Standard Reference Data System	NSRDS
natural	nat
natural logarithm	ln
natural logarithm	log$_e$
natural logarithm [British]	ln
natural logarithm [British]	log$_e$
natural logarithm [Canadian]	ln
natural logarithm [Canadian]	log$_e$
natural logarithm [ISO]	ln
natural logarithm [ISO]	log$_e$
natural logarithm [IUPAP]	ln

negative	neg
negative-on-positive (solar cell)	N-on-P
neodymium	Nd
neon	Ne
neper	Np
neper [British]	N
neper [ISO]	Np
neptunium	Np
network	net.
neutral	neut
newton	N
newton [British]	N
newton [IUPAC]	N
newton [IUPAP]	N
nickel	Ni
niobium (columbium)	Nb
nit	nt
nit [British]	nt
nitrogen	N
nobelium	No
noise reduction	NR
noise reduction coefficient	NRC
nonparametric	**NP**
nonreflected-shock tunnel	NRS
normal	norm.
normal (concentration)	N
normal (concentration) [British]	N
normal (concentration) [IUPAC]	N
normal temperature and pressure	NTP
north	N
northeast	NE
north-northeast	NNE
north-northwest	NNW
north-south	N-S
northwest	NW
nota bene (mark well)	N.B.
noun	n.
November	Nov.
nuclear	nucl
nuclear detonation	nudet
nuclear magnetic resonance	NMR
nuclear magneton	n.m.

nuclear power demonstration reactor	NPD
Nuclear Science Abstracts (of AEC)	NSA
number	no.
numeral	num
numerator	num
octagon	oct
numerical aperture	N.A.
numero (number)	No.
object	obj
objective	obj
observation	obs
observed	obs
observed [British]	obs.
obtuse	obt
obverse	obv
oceanography	oceanog
octahedral	octahdr
octal	oct
October	Oct.
oersted	Oe
oersted [AAAS]	oer
oersted [IUPAP]	Oe
ohm	Ω
ohm [British]	Ω
ohm [Canadian]	Ω
ohm [IUPAC]	Ω
ohm [IUPAP]	Ω
ohmmeter	ohm.
one-way	1/W
operator	oper
opere citato (in the work cited)	op. cit.
opposite	opp
optics	opt
orbit	orb.
orbiting astronomical observatory	OAO
orbiting geophysical observatory	OGO
orbiting planetary observatory	OPO
orbiting solar observatory	OSO
ordinal	ord
ordinary	ord
organic	org
organic-moderated reactor	OMR

orientation	**orient.**
origin	**orig**
orthogonal	**orthog**
oscillate	**osc**
osmium	**Os**
ounce	**oz**
ounce [British]	**oz**
ounce [Canadian]	**oz**
ounce, avoirdupois	**oz avdp**
ounce-foot	**oz-ft**
ounce-foot [Canadian]	**oz-ft**
ounce-force [British]	**ozf**
ounce-inch	**oz-in.**
ounce-inch [Canadian]	**oz-in.**
ounce, troy	**oz t**
outside diameter	**o.d.**
outside radius	**o.r.**
over-all	**OA**
overvoltage	**ovv**
oxygen	**O**
page	**p.**
pages	**pp.**
pair	**pr**
palladium	**Pd**
parabola	**parab**
paragraph	**par.**
parallax second	**parsec**
parallel	**par.**
parameter	**param**
parametric	**param**
parentheses	**parens**
parsec	**pc**
part	**pt**
partial vapor pressure	**pvp**
participle	**part.**
particle	**part.**
parts per billion	**ppb**
parts per million	**ppm**
parts per million [Canadian]	**ppm**
passed	**P**
Penning ionization gage	**PIG**
pentagon	**pent.**

percent	pct
perfect	perf
period	pd
permanent magnet	pm
permutation	perm
perpendicular	perp
perpendicular bisector	perp bis
perspective	persp
phase	ph
Philips ionization gage	PIG
phosphorus	P
phot	ph
physical	phys
physics	phys
pico (=10^{-12}) (prefix) [International]	p
picocurie	pc
picofarad	pf
picofarad [AIP]	pF
picofarad [IUPAP]	pF
picosecond	psec
pi-meson	pion
pint	pt
pint [Canadian]	pt
plane	pl
plate	pl
platinum	Pt
platinum-cobalt (color method)	Pt-Co
platinum resistance thermometer	prt
plenum	pln
plural	pl.
plutonium	Pu
plutonium recycle test reactor	PRTR
point	pt
point of closest approach	PCA
point of inflection	p.i.
poise	P
poise [British]	P
poise [IUPAC]	P
poise [IUPAP]	P
polar-cap absorption	PCA
polarity	pol
polar orbiting geophysical observatory	POGO

Polish	Pol.
polonium	Po
polygon	p-gon
polyhedron	p-hed
polynomial	p-nom
pool test reactor	PTR
Portuguese	Port.
positive	pos
positive-on-negative (solar cell)	P-on-N
post meridiem (afternoon)	p.m.
postulate	post.
potassium	K
potential difference	**PD**
potential difference [British]	**p.d.**
potentiometer	pot.
pound	lb
pound [British]	lb
pound [Canadian]	lb
poundal	pdl
poundal [British]	pdl
pound, avoirdupois	lb avdp
pound-foot	lb-ft
pound-foot [Canadian]	lb-ft
pound-force	lbf
pound-force [British]	lbf
pound-force-foot	lbf-ft
pound-force per square inch	lbf/in.²
pound-inch	lb-in.
pound-inch [Canadian]	lb-in.
pound-mass	lbm
pounds per cubic foot	pcf
pounds per cubic foot [ASA]	lb per cu ft
pounds per cubic foot [Canadian]	lb per cu ft
pounds per square foot	psf
pounds per square foot [Canadian]	psf
pounds per square inch	psi
pounds per square inch [Canadian]	psi
pound, troy	lb t
power demonstration reactor	PDR
power series	PS
praseodymium	Pr
precipitate	ppt

precipitate [British]	ppt.
preposition	prep.
pressure	press.
pressure gradient force	PGF
pressure wave	P-wave
pressurized water reactor	PWR
principal distance	P.D.
principal value of inverse cosecant	Arc csc
principal value of inverse cosine	Arc cos
principal value of inverse cotangent	Arc cot
principal value of inverse cotangent	Arc ctn
principal value of inverse secant	Arc sec
principal value of inverse sine	Arc sin
principal value of inverse tangent	Arc tan
prism diopter	pd
prisms	pr
probability	prob
probable error	PE
probable error [AIP]	pe
probes for the International Quiet Solar Year	PIQSY
problem	prob
process	proc
product	prod.
Professor	Prof.
program	prog
project	proj
projection	proj
promethium	Pm
pronoun	pron.
proportion	pptn
proportional	pptnl
proposition	prop.
protactinium	Pa
pro tempore (temporarily)	pro tem.
proton linear accelerator	PLA
proton synchrotron	PS
prototype organic power reactor	POPR
publication	publ
pulley	pul
pulse-amplitude modulation	PAM
pulse-code modulation	PCM
pulse-count modulation	PCM

pulsed ion linear accelerator	**PILAC**
pulse-duration modulation	**PDM**
pulse-frequency modulation	**PFM**
pulse modulation	**PM**
pulse-position modulation	**PPM**
pulses per minute	**ppm**
pulses per second	**pps**
pulse-time modulation	**PTM**
pulse-width modulation	**PWM**
pyramid	**pyr**
pyrometer	**pyr**
quadrangle	**quad**
quadrant	**quad**
quadratic	**quad**
quadrilateral	**quad**
quae vide (which see) (plural)	**qq.v.**
quality factor (radiation)	**QF**
quantum	**quant**
quantum amplification by stimulated emission of radiation	**QUASER**
quantum mechanics	**quant mech**
quantum number	**quant no.**
quart	**qt**
quart [Canadian]	**qt**
quarter	**qtr**
quartile deviation	**QD**
quasi-longitudinal	**QL**
quasi-transverse	**QT**
quod erat demonstrandum (which was to be demonstrated, or proved)	**Q.E.D.**
quod erat faciendum (which was to be done)	**Q.E.F.**
quod vide (which see)	**q.v.**
quotient	**quot**
radian	**rad**
radian [British]	**rad**
radian [ISO]	**rad**
radians per second [British]	**rad/s**
radians per second [ISO]	**rad/s**
radiant	**rad**
radiation	**radn**
radiation-thermal cracking	**RTC**
radical	**rad**

radioactivity concentration guides	RCG
radio detection and ranging	radar
radio-frequency (adjective)	rf
radiography	radiog
radium	Ra
radius	rad
radon	Rn
random variable	r.v.
range	R
rate of loss of energy	RLE
ratio	r
rational	ratnl
Rayleigh wave	R-wave
reactive kilovolt-ampere; kilovar	kvar
reactive kilovolt-ampere [Canadian]	kvar
reactive volt-ampere	var
reactive volt-ampere [Canadian]	var
reactivity	reac
reactivity coefficient	RC
reactivity measurement facility	RMF
reactor	reac
real part of	Re
real part of [British]	Re
real part of [ISO]	Re
real part of [IUPAP]	Re
recapitulation	recap
reciprocal	recip
reciprocal meter [ISO]	m^{-1}
reciprocal minute [ISO]	min^{-1}
reciprocal second [ISO]	s^{-1}
reciprocal second squared [ISO]	s^{-2}
rectangle	rect
reduce	red.
reduction-oxidation	redox
reference	ref
reflection	refl
refraction	refr
refractive index	RI
regarding	re
region	reg
regular	reg
relative	rel

relative biological effectiveness	**RBE**
relative ionospheric opacity meter	**riometer**
remainder	**rem**
report	**rept**
reproduction	**repro**
repulsion	**rep**
research	**res**
research and development	**R&D**
residual	**resid**
resistance-capacitance	**RC**
resistor	**R**
resistor	**res**
respectively	**resp**
rest mass unit	**rmu**
reverse	**rev**
review	**rev**
revolution	**rev**
revolutions per minute	**rpm**
revolutions per minute [British]	**rev/min**
revolutions per minute [Canadian]	**rpm**
revolutions per second	**rps**
revolutions per second [Canadian]	**rps**
rhenium	**Re**
rhodium	**Rh**
right	**rt**
right angle	**rt ang**
right ascension	**RA**
right ascension of mean sun	**RAMS**
rocket	**rkt**
Rockwell hardness	**Rh**
roentgen	**r**
roentgen [AIP]	**R**
roentgen [British]	**r**
roentgen equivalent, man	**rem**
roentgen equivalent, physical	**rep**
roentgens per hour	**r/hr**
root-mean-square	**rms**
root-mean-square [British]	**r.m.s.**
root-mean-square [Canadian]	**rms**
root-sum-square	**rss**
rotate	**rot.**
rotation	**rot**

Royal	Roy.
rubidium	Rb
rule of thumb	ROT
Russian	Russ.
ruthenium	Ru
rydberg	Ry
samarium	Sm
sample	samp
satellite	satel
saturate	sat.
saturated vapor pressure	svp
saturation	satn
Saturday	Sat.
scales	sc
scandium	Sc
scatter-specular ratio	cpr
schedule	sch
science	sci
scientific applications of nuclear explosions	SANE
scientist	sci
secant	sec
secant [British]	sec
secant [Canadian]	sec
secant [ISO]	sec
secant [IUPAP]	sec
second	sec
second [Canadian]	sec
second [ISO]	s
second [IUPAC]	s
second [IUPAP]	s
second (time) [British]	s
second (time, in astronomical tables)	s
second (time, in astronomical tables) [Canadian]	s
second sine-integral function	si
section	sect.
Section (in references)	Sec.
sector	sect.
segment	seg
selenium	Se
self-consistent field (calculations)	SCF
self-resonant frequency	SRF
semidiameter	sd

September	Sept.
sequence	seq
shear wave	S-wave
short range	SR
short wave	SW
side, angle, side	s.a.s.
sidereal hour angle	SHA
side, side, side	s.s.s.
signal-noise ratio	SNR
signal-noise ratio [AGU]	snr
silicon	Si
silver	Ag
similar	sim
simple harmonic motion	SHM
simultaneous	simul
simultaneous equations	simul eqs
sine	sin
sine [British]	sin
sine [Canadian]	sin
sine [ISO]	sin
sine [IUPAP]	sin
sine of the amplitude, an elliptic function	sn
sine of the amplitude, an elliptic function [Canadian]	sn
Smithsonian Astrophysical Observatory	SAO
sodium	Na
sodium graphite reactor	SGR
solar	sol
Solar Energy Thermionic (program)	SET
Solar Energy Thermionic Conversion System	SETS
solid	sol
solvent	solv
sound	snd
sound pressure level	SPL
sound velocity, temperature, and pressure	SVTP
south	S
southeast	SE
south-southeast	SSE
south-southwest	SSW
southwest	SW
Spanish	Span.
specific activity	sp act.

specification	spec
specific gravity	sp gr
specific gravity [British]	sp.gr.
specific gravity [Canadian]	sp gr
specific heat	sp ht
specific heat [British]	sp.ht.
specific heat [Canadian]	sp ht
specific inductive capacity	sic
specific volume	sp vol
sphere	sph
spherical	spher
spherical candlepower	scp
spherical candlepower [Canadian]	scp
square	sq
square [Canadian]	sq
square centimeter	cm^2
square centimeter	sq cm
square centimeter [British]	cm^2
square centimeter [Canadian]	cm^2
square centimeter [Canadian]	sq cm
square decimeter	dm^2
square foot	ft^2
square foot [British]	ft^2
square foot [Canadian]	sq ft
square foot [ISO]	ft^2
square inch	$in.^2$
square inch [British]	in^2
square inch [Canadian]	sq in.
square inch [ISO]	in^2
square meter	m^2
square meter [British]	m^2
square meter [Canadian]	m^2
square meter [Canadian]	sq m
square meter [ISO]	m^2
square micron	μ^2
square micron [Canadian]	sq mu
square micron [Canadian]	sq μ
square micron [Canadian]	μ^2
square millimeter	mm^2
square millimeter [Canadian]	mm^2
square millimeter [Canadian]	sq mm
square root	sq rt

square yard	yd²
square yard [ISO]	yd²
stable	stab.
standard	std
standard [Canadian]	std
standard deviation	SD
standard error	SE
standard temperature and pressure	STP
standard temperature and pressure [British]	s.t.p.
standing-wave ratio	swr
station	sta
statistical	stat
statistician	statist
steradian	sr
steradian [ISO]	sr
stilb	sb
stilb [British]	sb
stochastic	stoch
stochastic process method	SPM
stoichiometric	stoich
stoke	S
stoke [British]	S
straight	str
strength	str
stress to number of cycles curve	S-N curve
strontium	Sr
sublimes	subl
submarine	sub
substitute	sub
substitution	sub.
subtraction	subtr
sub verbo (under the word)	s.v.
sub voce (under the word)	s.v.
sudden auroral intensity	SAI
sudden cosmic noise absorption	SCNA
sudden enhancement of atmospherics	SEA
sudden ionospheric disturbance	SID
sudden short-wave fade	**S-SWF**
sulfur	S
sum of squares	**SS**
Sunday	Sun.
super-high frequency	shf

supplement	suppl
supremum	sup
suspension	susp
Swedish	Swed.
switch	sw
symbol	sym
symmetrical	sym
symmetrically cyclically magnetized (condition)	SCM
symposium	symp
synonym	syn.
synthesis	synth
tabulate	tab.
tachometer	tach
tangent	tan
tangent [British]	tan
tangent [Canadian]	tan
tangent [ISO]	tan
tangent [ISO]	tg
tangent [IUPAP]	tan
tangent [IUPAP]	tg
tantalum	Ta
target	tgt
technetium	Tc
technical report	TR
technological	technol
telescope	tel
tellurium	Te
temperature	temp
temperature [British]	temp.
temperature [Canadian]	temp
temperature-humidity index	THI
tensile strength	ts
tensile strength [Canadian]	ts
tensile yield strength	TYS
Ten-Year Oceanographic Program	TENOC
tera ($=10^{12}$) (prefix) [International]	T
teracycle	Tc
terbium	Tb
terrestrial north pole	TNP
terrestrial principal point	TPP
terrestrial spin-axis point	TSA
terrestrial subpoint	TSP

tesla	T
tesla [IUPAP]	T
tetragonal	tetr
tetrahedron	tetrah
thallium	Tl
the following (in citations)	f.
the following (plural) (in citations)	ff.
theorem	Th
theoretical mineral acidity	TMA
theory	th
thermal efficiency	TE
thermodynamics	thermodyn
thermogravimetric analysis	TGA
thermogravimetric analyzer	TGA
thermomagnetic treatment	TT
thermometer	therm
thermoremanent magnetization	TRM
thin-layer chromatography	TLC
thin-layer electrophoresis	TLE
thin region integral method	TRIM
thorium	Th
thousand	M
thousand [Canadian]	M
three-dimensional	3-D
threshold limit value	TLV
thulium	Tm
Thursday	Thurs.
time	t
time-reversal invariant	TRI
tin (stannum)	Sn
titanium	Ti
tolerance	tol
ton-force [British]	tonf
tonne (1000 kilograms)	t
tonne (1000 kilograms) [British]	t
tonne (1000 kilograms) [IUPAC]	t
tonne (1000 kilograms) [IUPAP]	t
torsion	tor
trajectory	traj
transequatorial scatter	TE
transformation	xform
transistor	xstr

transition point	tr
transitive	tr.
transit research and altitude control (satellite)	TRAAC
translation	transl
transmission loss	TL
transversal	transv
transverse	transv
transverse electric	TE
transverse electromagnetic	TEM
transverse electromagnetic wave	TEM wave
transverse magnetic	TM
transverse magnetic wave	TM wave
trapezoid	trap.
traveling ionospheric disturbance	TID
traveling-wave tube	TWT
triclinic	tricl
trigonal	trig
trigonometry	trig
tritium	T
tritium-hydrogen ratio	T/H
troy	t
true vapor pressure	tvp
true watt	tw
Tuesday	Tues.
tungsten	W
two-dimensional	2-D
two-way	2/W
ubi supra (in the place above mentioned)	u.s.
ultra-high frequency	uhf
ultra-short wave	USW
ultrasonic frequency	uf
ultraviolet	uv
ultraviolet [British]	u.v.
ultraviolet amplification by stimulated emission of radiation	UVASER
Union of Soviet Socialist Republics	USSR
unitary complex	u.c.
United Kingdom	UK
United States Antarctic Research Program	USARP
units-of-variance	UOV
universal automatic computer	Univac
Universal Decimal Classification	UDC

universe	**univ**
university	**univ**
university training reactor	**UTR**
unknown	**unk**
up	**U**
upper yield point	**UYP**
uranium	**U**
vacuum	**vac**
vacuum [British]	**vac.**
vacuum tube	**VT**
value	**val**
vanadium	**V**
vapor	**vap**
vapor density	**vd**
vapor density [British]	**v.d.**
vapor pressure	**vp**
vapor pressure [British]	**v.p.**
variable	**var**
variance	**var**
velocity	**vel**
velocity gravity constant	**vgc**
velocity-of-propagation meter	**VP-meter**
verb	**v.**
versed sine	**vers**
versed sine [Canadian]	**vers**
very-high frequency	**vhf**
very-low frequency	**vlf**
vibrating-coil magnetometer	**VCM**
vibrations per minute	**vpm**
vibrations per second	**vps**
Vickers hardness number	**Vhn**
Vickers pyramid number	**Vpn**
vide (see)	**v.**
videlicet (namely)	**viz.**
vide supra (see above)	**v.s.**
viscosity	**visc**
viscosity index	**V.I.**
volt	**v**
volt [AIP]	**V**
volt [British]	**V**
volt [Canadian]	**v**

volt [IUPAC]	V
volt [IUPAP]	V
voltage standing-wave ratio	vswr
volt-ampere [British]	VA
volt-ampere [Canadian]	va
volt-ampere [LRL]	VA
volt-coulomb [British]	VC
voltmeter	VM
volume	vol
volume [British]	vol.
watt	w
watt [AIP]	W
watt [British]	W
watt [Canadian]	w
watt [IUPAC]	W
watt [IUPAP]	W
watt-hour	Wh
watt-hour [British]	Wh
watt-hour [Canadian]	whr
watt-hour [LRL]	W-h
wavelength	wl
weber	Wb
weber [British]	Wb
weber [IUPAP]	Wb
Wednesday	Wed.
weight	wt
weight [British]	wt.
weight [Canadian]	wt
weight percent	w/o
well-known fact	WKF
west	W
west-northwest	WNW
west-southwest	WSW
wet bulb	WB
window atmospheric sounding projectile	WASP
World Magnetic Survey	WMS
xenon	Xe
x-unit	xu
yard	yd
yard [British]	yd
yard [Canadian]	yd
yard [ISO]	yd

year .... yr ....
year [Canadian] .... yr ....
year [ISO] .... a ....
year [IUPAP] .... a ....
yield point .... YP ....
yield strength .... YS ....
ytterbium .... Yb ....
yttrium .... Y ....
zenith distance .... ZD ....
zero frequency .... ZF ....
zero gradient synchrotron .... ZGS ....
zinc .... Zn ....
zirconium .... Zr ....
zone time .... ZT ....

LETTER SYMBOLS

A LETTER SYMBOL denotes a scalar or a physical quantity or a mathematical statement and equation. It is usually a single character of the English or Greek alphabet, with modifying subscript or superscript if necessary.

Units with names or symbols are used to express physical quantities, whereas the units for dimensionless quantities have no name or symbol, and as such, are not explicitly indicated.

In practice, the same symbol should designate the same physical magnitude regardless of the units used and special values occurring for different conditions.

In practice italic type is preferred for letter symbols. For subscripts and superscripts, American usage prescribes italics while most international sources prefer roman type. Abbreviations used as subscripts are set in roman type. Vectors are usually printed in boldface type, but sometimes an italic letter with an arrow over it is also used. Ordinary Hindu numerals are prescribed where numbers must be stated as coefficients, exponents, *etc.*

Much duplication in symbology results from the inadequate supply of letters forced to represent the entire mass of functions and parameters considered in practice. Alternate symbols for many quantities are, therefore, listed to minimize repetition.

The specificity and transposed forms of terminology as reported by the sources have been maintained and listed alphabetically throughout this section. For example, both "power, reactive" [British] and "reactive power" [IEC] definitions for the symbol "Q" are listed.

Subscripts and superscripts have been included in the major list with parenthetical notes explaining their use.

Entries without key signatures are from recognized American sources. Conflicting symbols, and entries from foreign or international sources, are keyed as follows:

[ASA] .... American Standards Association
[British] .... indicates symbols recommended by one or both of the following: British Standards Institution or British Institution of Electrical Engineers

DPMA

[IEC] . . . . .International Electrotechnical Commission
[ISO] . . . . .International Organization for Standardization
[IUPAC] . . . . . . .International Union of Pure and Applied Chemistry
[IUPAP] . . . . . . .International Union of Pure and Applied Physics

LETTER SYMBOLS
Alphabetically by Symbol

0	....	neutral charge (superscript) [IUPAP] ....
0	....	reference state (subscript) [British] ....
0	....	standard or limiting condition (subscript) [British] .
a	....	absolute value (subscript) ....
a	....	absorption coefficient [British] ....
a	....	absorption coefficient [IUPAP] ....
a	....	absorptivity [IUPAC] ....
a	....	acceleration ....
a	....	acceleration [ISO] ....
a	....	acceleration [IUPAC] ....
a	....	acceleration [IUPAP] ....
a	....	activity, chemical ....
a	....	adsorbed (subscript) ....
a	....	ambient (subscript) ....
a	....	aperture ....
a	....	arithmetic (subscript) ....
a	....	attenuation coefficient [IEC] ....
a	....	coefficient of accommodation ....
a	....	dry adiabatic processes (subscript) ....
a	....	Helmholtz function, specific ....
a	....	interfacial area per common volume [British] .....
a	....	linear acceleration [British] ....
a	....	linear acceleration [IEC] ....
a	....	radius of a diaphragm ....
a	....	radius of a disk ....
a	....	radius of a membrane ....
a	....	radius of a tube ....
a	....	radius of Earth ....
a	....	relative activity [British] ....
a	....	relative activity [IUPAC] ....
a	....	slit width ....
a	....	specific absorbance [IUPAC] ....
a	....	speed of sound [British] ....
a	....	surface area per unit volume ....
a	....	thermal diffusivity [British] ....
a	....	total acoustical absorption in a room ....
a_0	....	Bohr radius [IUPAP] ....
a_1	....	Bohr radius ....

a_B	relative activity of substance B [IUPAP]
a_m	Helmholtz function per atom or molecule
a_m	maximum isothermal work function per atom or molecule
a_x	relative activity of substance X [British]
ad	adiabatic (subscript)
aw	adiabatic wall (subscript)
a	linear acceleration
A	absorbance (light) [IUPAC]
A	activity [IUPAP]
A	affinity
A	affinity [IUPAP]
A	affinity of a chemical reaction [IUPAC]
A	albedo
A	amplification of amplifier, power
A	amplification of amplifier, voltage
A	amplitude
A	amplitude of velocity potential
A	angle of prism, refracting
A	anode terminal
A	area
A	area [IEC]
A	area [ISO]
A	area [IUPAC]
A	area [IUPAP]
A	aspect ratio
A	aspect ratio [British]
A	atomic weight
A	atomic weight [British]
A	Austausch coefficient
A	Cauchy constant
A	component A (subscript)
A	cross-sectional area
A	cross-sectional area [British]
A	dynamic coefficient of eddy transfer
A	extinction (light) [IUPAC]
A	free energy, Helmholtz function
A	gain of amplifier, power
A	gain of amplifier, voltage
A	heat equivalent of work
A	Helmholtz free energy [IUPAC]
A	Helmholtz function
A	linear current density

DPMA

A	linear current density [IEC]	
A	magnetic vector potential [British]	
A	major hydraulic area	
A	mass number	
A	mass number [IUPAP]	
A	maximum isothermal work function	
A	recombination coefficient [IUPAP]	
A	reference area for drag and lift	
A	strength of a simple source	
A	surface area	
A	work [IEC]	
A	work [IUPAC]	
A	work [IUPAP]	
$\tilde{A}$	transpose of matrix A [IUPAP]	
A^*	complex conjugate of A [IUPAP]	
$A^\dagger$	Hermitian conjugate of A [IUPAP]	
A_p	amplification of amplifier, power	
A_p	gain of amplifier, power	
A_r	relative atomic mass [IUPAP]	
A_v	amplification of amplifier, voltage	
A_v	gain of amplifier, voltage	
A_v	surface area per unit volume	
$\mathbf{A}$	axial vector coupling [IUPAP]	
$\mathbf{A}$	vector potential [IUPAP]	
$\mathbf{A}$	magnetic vector potential	
b	base (subscript)	
b	black body (subscript)	
b	breadth	
b	breadth [British]	
b	breadth [IEC]	
b	breadth [ISO]	
b	breadth [IUPAP]	
b	impact parameter [IUPAP]	
b	phase coefficient [IEC]	
b	semichord	
b	semiminor axis	
b	susceptance	
b	width	
b	Wien's displacement constant	
b_A	acoustical susceptance	
b_E	electrical susceptance	
b_M	mechanical susceptance	
b_R	rotational susceptance	

bar	barometric (subscript)	
B	black body (subscript)	
B	brightness	
B	Cauchy constant	
B	component B (subscript)	
B	compressibility factor, supersonic [British]	
B	flux density	
B	luminance	
B	luminance [IUPAC]	
B	magnetic flux density [British]	
B	magnetic flux density [IEC]	
B	magnetic induction [British]	
B	magnetic induction [IEC]	
B	magnetic induction [IUPAC]	
B	magnetic induction [IUPAP]	
B	photometric brightness	
B	second virial coefficient	
B	susceptance	
B	susceptance [British]	
B	susceptance [IEC]	
B	susceptance [IUPAP]	
B	volume modulus of elasticity	
$\bar{B}$	average binding energy per nucleon in a nucleus [British]	
B_A	acoustical susceptance	
B_E	electrical susceptance	
B_i	intrinsic induction	
B_M	mechanical susceptance	
B_R	rotational susceptance	
[B]	molar concentration of substance B [IUPAC]	
B	magnetic induction [IUPAP]	
B	flux density, magnetic	
B	induction density, magnetic	
$\mathcal{B}$	steradiance	
c	acoustic conductivity of an opening	
c	chord	
c	chord [British]	
c	chord (subscript)	
c	cloud (subscript)	
c	coefficient	
c	coefficient of aging (piezoelectricity)	
c	concentration	

c	concentration [British]
c	concentration [IUPAC]
c	condensation level (subscript)
c	condensation processes (subscript)
c	critical properties (subscript)
c	critical state (subscript)
c	critical value (subscript)
c	damping coefficient, velocity
c	distance to extreme fiber from neutral axis
c	elastic stiffness
c	factor
c	heat capacity per mole
c	heat capacity per unit mass
c	heat capacity, specific [British]
c	induction coefficient
c	mass concentration
c	partial capacitance coefficient
c	permittance
c	specific heat [IEC]
c	specific heat capacity
c	speed of light [British]
c	speed of light in empty space [IUPAP]
c	speed of light in vacuum
c	speed of light in vacuum [ISO]
c	speed of sound
c	velocity [ISO]
c	velocity [IUPAP]
c	velocity of light in vacuum
c	velocity of light in vacuum [IEC]
c	velocity of sound [IUPAP]
c	wave velocity
$\bar{c}$	average velocity [IUPAP]
$\hat{c}$	most probable speed [IUPAP]
c_0	speed, most probable
c_1	Planck's radiation law constant
c_2	Planck's radiation law constant
c_{33}	elastic stiffness constant
c_B	molar concentration of substance B
c_B	molar concentration of substance B [IUPAC]
c_B	molar concentration of substance B [IUPAP]
c_f	surface friction coefficient, local [British]
c_g	group velocity [IUPAP]

c_l .... velocity of longitudinal waves [IUPAP] ....
c_m .... heat capacity per atom ....
c_m .... heat capacity per molecule ....
c_p .... **specific heat at constant pressure [IUPAC]** ....
c_p .... specific heat capacity [IUPAP] ....
c_p .... specific heat capacity at constant pressure ....
c_p .... specific heat capacity at constant pressure [British] .
c_t .... velocity of transverse waves [IUPAP] ....
c_v .... specific heat at constant volume [IUPAC] ....
c_v .... specific heat capacity [IUPAP] ....
c_v .... specific heat capacity at constant volume ....
c_v .... specific heat capacity at constant volume [British] ..
c_x .... molar concentration of substance X [British] ......
$c(B)$ .... molar concentration of substance B [IUPAC] .....
c .... convection (subscript) [British] ....
c .... critical state (subscript) [British] ....
c .... critical value (subscript) [British] ....
cal .... calibrated (subscript) ....
calc .... calculated (subscript) ....
$\cos \phi$ .... power factor (sinusoidal quantities) [British] .....
c .... molecular velocity vector [IUPAP] ....
$\mathbf{c}_0$ .... average velocity [IUPAP] ....
$\mathbf{c}_0$ .... velocity, most probable ....
C .... capacitance ....
C .... capacitance [British] ....
C .... capacitance [IEC] ....
C .... capacitance [IUPAP] ....
C .... capacity [IUPAC] ....
C .... Cauchy constant ....
C .... Chezy coefficient ....
C .... Chezy coefficient [British] ....
C .... circulation ....
C .... coefficient ....
C .... coefficient [British] ....
C .... compliance ....
C .... component C (subscript) ....
C .... concentration [British] ....
C .... constant ....
C .... Curie constant ....
C .... factor ....
C .... heat capacity [IUPAC] ....
C .... heat capacity per mole ....

C	heat capacity per mole [British]	
C	molecular concentration [IUPAC]	
C	normality	
C	permittance	
C	strength of a vortex	
C	thermal conductance	
C	total heat capacity	
C_A	acoustical capacitance	
C_D	coefficient of drag	
C_D	coefficient of drag [British]	
C_E	electrical capacitance	
C_f	surface friction coefficient, over-all [British]	
C_M	heat capacity per mole	
C_M	mechanical compliance	
C_p	heat capacity at constant pressure	
C_p	heat capacity at constant pressure [British]	
C_p	molar heat capacity [IUPAP]	
C_p	coefficient of pressure [British]	
C_R	rotational compliance	
C_v	heat capacity at constant volume	
C_v	heat capacity at constant volume [British]	
C_v	molar heat capacity [IUPAP]	
C_X	molar concentration of substance X [British]	
d	dew point (subscript)	
d	diameter	
d	diameter [British]	
d	diameter [IEC]	
d	diameter [ISO]	
d	diameter [IUPAC]	
d	diameter [IUPAP]	
d	differential operator	
d	differential operator [IEC]	
d	distance	
d	distance between corresponding points of grating	
d	distance between lens units in an optical system	
d	distance from focus to directrix of conic section	
d	dry air (subscript)	
d	interplanar distance (Bragg law)	
d	optical density [British]	
d	piezoelectric strain constant	
d	relative density [British]	
d	relative density [IUPAC]	
d	relative density [IUPAP]	

d	spacing of Bragg planes in a crystal	
d	thickness [ISO]	
d_h	hydrostatic piezoelectric strain constant	
d_p	planar piezoelectric strain constant	
d	deuteron	
d	deuteron [IUPAP]	
d	dilution (subscript) [British]	
d	dissolution (subscript) [British]	
dx	total differential of x [IUPAP]	
D	angle of minimum deviation	
D	angular dispersion	
D	coefficient of diffusion [IUPAC]	
D	constant electric displacement (superscript)	
D	declination of geomagnetic field	
D	density	
D	diameter	
D	diameter [British]	
D	diameter of molecule [IUPAC]	
D	diffusion coefficient [British]	
D	diffusion coefficient [IUPAP]	
D	dioptric power	
D	dissipation factor	
D	drag	
D	drag [British]	
D	drag (subscript)	
D	duration of rainfall	
D	electric displacement [British]	
D	electric displacement [IUPAC]	
D	electric displacement [IUPAP]	
D	electric flux density [British]	
D	electric flux density [IEC]	
D	fourth virial coefficient	
D	mass diffusion coefficient	
D	mass diffusivity	
D	mass per unit volume	
D	optical attenuation	
D	optical density	
D	power of lens system	
D	rainfall duration	
D	refracting power	
D_0	remanent or bias displacement	
D_i	directivity index	
D_T	thermal diffusion coefficient	

D_T		thermal diffusion coefficient [IUPAP]
D_v		volumetric diffusivity
D		electric displacement [IUPAP]
D		dielectric flux density
D		displacement flux density
D		electric displacement
D		electric flux density
D		induction density, electric
$\dot{D}$		displacement current
$\dot{D}/4\pi$		displacement current
e		base of Naperian logarithms
e		base of natural logarithms [IEC]
e		charge of positron [IUPAP]
e		coefficient of resilience
e		coefficient of restitution
e		eccentricity
e		eccentricity [British]
e		elementary charge [IUPAC]
e		energy of gas, specific internal [British]
e		equivalent quantities (subscript)
e		external (subscript)
e		piezoelectric stress constant
e		water vapor pressure
$-e$		charge of electron [British]
$-e$		electronic charge
e		base of natural logarithms [British]
e		base of natural logarithms [IUPAP]
e		electric (subscript) [IUPAP]
e		electron
e		electron [IUPAP]
e		evaporation (subscript) [British]
e		extrapolated value of a length (subscript) [British]
e^x		exponential of x [IUPAP]
e		elementary charge [IUPAC]
E		absorbance (light) [IUPAC]
E		activation energy [British]
E		amount of illumination
E		constant electric field (superscript)
E		constant energy (subscript)
E		electric field [IUPAP]
E		electric field strength [British]
E		electric field strength [IEC]

DPMA

E	....	electric field strength [IUPAC] ....
E	....	electric force [British] ....
E	....	electric intensity [IEC] ....
E	....	electromotive force ....
E	....	electromotive force [British] ....
E	....	electromotive force [IEC] ....
E	....	electromotive force [IUPAC] ....
E	....	energy ....
E	....	energy [British] ....
E	....	energy [IEC] ....
E	....	energy [IUPAC] ....
E	....	energy [IUPAP] ....
E	....	energy density ....
E	....	energy of gas, internal [British] ....
E	....	energy, total ....
E	....	extinction (light) [IUPAC] ....
E	....	illuminance ....
E	....	illumination [IUPAC] ....
E	....	illumination [IUPAP] ....
E	....	modulus of elasticity [IUPAC] ....
E	....	modulus of elasticity [IUPAP] ....
E	....	potential difference ....
E	....	voltage ....
E	....	voltage, maximum ....
E	....	Young's modulus [British] ....
E	....	Young's modulus [IUPAP] ....
E	....	Young's modulus of elasticity ....
E_{co}	....	cutoff voltage ....
E_e	....	irradiance [IUPAP] ....
E_f	....	filament supply voltage, alternating-current ....
E_k	....	kinetic energy ....
E_k	....	kinetic energy [IUPAP] ....
E_p	....	potential energy ....
E_p	....	potential energy [IUPAP] ....
E_v	....	energy of vibration ....
$\mathbf{E}$	....	electric field [IUPAP] ....
$\mathbf{E}_h$	....	Hall field ....
$\mathbf{E}_h$	....	transverse electric field ....
$\mathbf{E}$	....	electric field strength ....
$\mathbf{E}$	....	electric gradient ....
$\mathbf{E}$	....	electric intensity ....
E	....	electromotive force ....

$\mathcal{E}$		irradiance
$\mathcal{E}$		radiant flux density
f		activity coefficient [British]
f		activity coefficient [IUPAC]
f		coefficient of friction
f		coefficient of sliding friction
f		Coriolis parameter
f		degrees of freedom
f		distribution function
f		ellipsoid flattening
f		filament (subscript)
f		final values (subscript)
f		focal length of object space
f		focus of conic section
f		free energy, specific [British]
f		frequency
f		frequency [British]
f		frequency [IEC]
f		frequency [ISO]
f		frequency [IUPAC]
f		frequency [IUPAP]
f		frequency (subscript)
f		friction coefficient [IUPAC]
f		friction coefficient [IUPAP]
f		friction factor
f		friction factor [British]
f		frost point (subscript)
f		fugacity
f		function
f		fusion processes (subscript)
f		linear acceleration [British]
f		normal stress [British]
f		packing fraction [IUPAP]
f		relative humidity
f'		focal length of image space
f_A		antiresonant frequency
f_r		natural frequency
f_r		resonant frequency
f_R		resonant frequency
$f(c)$		velocity distribution function [IUPAP]
$f(x)$		function of x [IUPAP]

f	function of [British]	
f	fusion (subscript) [British]	
F	area of equatorial plane of Earth	
F	**concentrated force**	
F	concentrated load	
F	distribution function	
F	energy fluence of particles	
F	**factor [British]**	
F	Faraday constant	
F	Faraday constant [British]	
F	Faraday constant [IUPAC]	
F	Faraday constant [IUPAP]	
F	fission rate	
F	force	
F	force [British]	
F	force [IEC]	
F	force [IUPAC]	
F	force [IUPAP]	
F	formality	
F	free energy, Gibbs function	
F	free energy, Helmholtz function [British]	
F	Froude number	
F	Froude number (fluid mechanics) [British]	
F	Gibbs function [IUPAC]	
F	gravitational force	
F	Helmholtz free energy [IUPAC]	
F	Helmholtz function [IUPAP]	
F	hyperfine quantum number [IUPAP]	
F	luminous flux	
F	luminous flux [British]	
F	magnetomotive force [British]	
F	magnetomotive force [IEC]	
F	quantum number, hyperfine	
F	stretching force in membrane	
F	stretching force in string	
F	tension in membrane	
F	tension in string	
F	thrust	
F	weight	
F_{10}	decimetric solar flux	
F_g	gravitational force	
F_h	horizontal force	

F_m	....	magnetomotive force [IEC] ....
F_r	....	resultant force ....
F_v	....	vertical force ....
Fr	....	Froude number [British] ....
$\mathbf{F}$	....	Faraday constant [IUPAC] ....
$\mathbf{F}$	....	force [IUPAP] ....
$\mathcal{F}$	....	magnetic scalar potential ....
$\mathcal{F}$	....	magnetomotive force ....
$\mathcal{F}$	....	magnetomotive force [IEC] ....
g	....	acceleration due to gravity ....
g	....	acceleration due to gravity [British] ...
g	....	acceleration due to gravity [IEC] ....
g	....	acceleration due to gravity [ISO] ....
g	....	acceleration due to gravity [IUPAP] ..
g	....	acceleration of free fall [IUPAC] ....
g	....	conductance ....
g	....	control grid (subscript) ....
g	....	degeneracy ....
g	....	geostrophic wind (subscript) ......
g	....	Gibbs function per mole ....
g	....	Gibbs function per unit mass ......
g	....	Gibbs function, specific ....
g	....	Gibbs function, specific [British] ....
g	....	group (subscript) ....
g	....	gyromagnetic ratio ....
g	....	Landé factor ....
g	....	osmotic coefficient [British] ....
g	....	osmotic coefficient [IUPAC] ....
g	....	osmotic coefficient [IUPAP] ....
g	....	slip [IEC] ....
g	....	statistical weight ....
g	....	statistical weight [IUPAC] ....
$-g$	....	acceleration of free fall ....
g_A	....	acoustical conductance ....
g_c	....	inertia proportionality factor ......
g_E	....	electrical conductance ....
g_g	....	grid conductance ....
g_g	....	input conductance ....
g_{gp}	....	inverse transconductance ....
g_L	....	local acceleration due to gravity ....
g_m	....	Gibbs function per atom ....
g_m	....	Gibbs function per molecule ....

g_m		grid-plate transconductance
g_m		mutual conductance [British]
g_M		mechanical conductance
g_*		inverse transconductance
g_n		standard acceleration due to gravity [British]
g_n		standard gravitational acceleration [IUPAP]
g_o		standard acceleration due to gravity
g_p		output conductance
g_p		plate conductance
g_{pg}		grid-plate transconductance
g_R		rotational conductance
gr		gradient wind (subscript)
g		gas (subscript) [IUPAP]
G		conductance
G		conductance [British]
G		conductance [IEC]
G		conductance [IUPAP]
G		force due to gravity [IEC]
G		force due to gravity [IUPAC]
G		free energy, Gibbs function
G		gas (subscript)
G		Gibbs function
G		Gibbs function [British]
G		Gibbs function [IUPAC]
G		Gibbs function [IUPAP]
G		Gibbs function per mole
G		Gibbs function, total value
G		gravitational constant [British]
G		gravitational constant, Newtonian
G		mass velocity
G		mass velocity [British]
G		modulus of rigidity [British]
G		shear modulus [British]
G		shear modulus [IUPAC]
G		shear modulus [IUPAP]
G		shear modulus of elasticity
G		vapor (subscript)
G		weight [IUPAC]
G		weight [IUPAP]
G_A		acoustical conductance
G_E		electrical conductance
G_M		Gibbs function per mole

G_M	mechanical conductance	
G_n	equivalent noise conductance	
G_R	rotational conductance	
G_z	system rating constant	
Gr	**Grashof number**	
Gr	Grashof number [British]	
Gz	Graetz number [British]	
G	gas (subscript) [British]	
h	altitude	
h	altitude (subscript)	
h	angular momentum, specific	
h	constant enthalpy (subscript) [British]	
h	degree of hydrolysis	
h	depth	
h	depth [IEC]	
h	enthalpy per atom or molecule	
h	enthalpy per mole	
h	enthalpy per unit mass	
h	head	
h	heat (subscript)	
h	heater (subscript)	
h	heat transfer coefficient	
h	heat transfer coefficient [British]	
h	height	
h	height [British]	
h	height [IEC]	
h	height [ISO]	
h	height [IUPAC]	
h	height [IUPAP]	
h	horizontal (subscript)	
h	horizontal surface (subscript)	
h	Planck's constant	
h	Planck's constant [British]	
h	Planck's constant [IEC]	
h	Planck's constant [IUPAC]	
h	Planck's constant [IUPAP]	
h	potential energy per unit weight	
h	radius of lens zone	
h	specific enthalpy	
h	specific enthalpy [British]	
h	surface heat transfer coefficient	
h	thickness	

h_{33}	piezoelectric stiffness constant	
h_f	friction head	
h_f	friction head [British]	
h_m	enthalpy per atom	
h_m	enthalpy per molecule	
h_{max}	maximum head [ASA]	
h_{max}	maximum head	
h_p	pressure head	
h_p	pressure head [British]	
h_v	velocity head	
h_v	velocity head [British]	
h	Planck's constant [IUPAC]	
H	angular momentum, total	
H	Boltzmann's function	
H	Boltzmann's function [IUPAP]	
H	boundary-layer shape parameter, in two-dimensional and axi-symmetric flow [British]	
H	depth	
H	enthalpy	
H	enthalpy [British]	
H	enthalpy [IUPAC]	
H	enthalpy [IUPAP]	
H	enthalpy per mole	
H	geopotential altitude	
H	Hamiltonian function	
H	Hamiltonian function [IUPAP]	
H	Hamiltonian operator	
H	heat content	
H	height of homogeneous atmosphere	
H	Henry law constant	
H	higher (subscript)	
H	horizontal thrust	
H	humidity	
H	irradiance	
H	luminous emittance [IUPAC]	
H	magnetic field [IUPAP]	
H	magnetic field strength	
H	magnetic field strength [British]	
H	magnetic field strength [IEC]	
H	magnetic field strength [IUPAC]	
H	magnetic intensity	
H	magnetic intensity [IEC]	
H	magnetizing force	

H	magnetizing force [British]	
H	power dissipated per unit volume (piezoelectricity)	
H	radiant flux density	
H	total enthalpy	
H	total heat [British]	
H_a	humidity at adiabatic saturation temperature	
H_a	humidity at adiabatic saturation temperature [British]	
H_c	coercive force	
H_d	dynamic height	
H_M	enthalpy per mole	
H_R	relative humidity	
H_R	relative humidity [British]	
H_s	humidity at saturation	
H_S	humidity at saturation [British]	
H_w	humidity at wet-bulb temperature	
H_w	humidity at wet-bulb temperature [British]	
H_λ	spectral irradiance	
H	magnetic field [IUPAP]	
H	magnetic field strength	
H	magnetic intensity	
H	magnetizing force	
$\mathcal{H}$	Hamiltonian function, perturbing	
i	angle of incidence	
i	current, instantaneous	
i	current, instantaneous [British]	
i	electric current [IUPAC]	
i	imaginary unit	
i	ice (subscript)	
i	imaginary unit	
i	imaginary unit [IUPAP]	
i	indicated (subscript)	
i	inertia of photographic plate	
i	initial value (subscript)	
i	input (subscript)	
i	inside (subscript)	
i	internal (subscript)	
i	mole factor	
i	order of overtone	
i	right-angle turning operator	
i	right-angle turning operator [IEC]	
i	square root of minus one	
i	Van't Hoff coefficient	

DPMA

i	vapor pressure constant
i_R	instantaneous current through resistance
$\mathbf{i}$	unit vector along X-axis
I	acoustic intensity
I	activity at specified time (radioactivity)
I	candlepower
I	conduction current
I	convection current
I	current
I	current [IEC]
I	current, root-mean-square
I	current, steady direct
I	direct solar radiant intensity, in or below atmosphere
I	effective current
I	electric current [British]
I	electric current [IUPAC]
I	electric current [IUPAP]
I	impulse
I	integration constant of Gibbs function equation
I	intensity
I	intensity of magnetization [British]
I	intensity of radiation
I	intensity of radiation [British]
I	ionic strength [British]
I	ionic strength [IUPAC]
I	ionic strength [IUPAP]
I	isotopic number [British]
I	luminous intensity
I	luminous intensity [British]
I	luminous intensity [IUPAC]
I	luminous intensity [IUPAP]
I	magnetic polarization [British]
I	magnetic shell strength
I	moment of inertia
I	moment of inertia [British]
I	moment of inertia [IEC]
I	moment of inertia [IUPAC]
I	moment of inertia [IUPAP]
I	moment of inertia, areal
I	nuclear spin quantum number
I	nuclear spin quantum number [IUPAP]
I	rectangular moment of inertia

I	resistivity index
I	second moment of area [British]
$\bar{I}$	average current
$\bar{I}$	quiescent current
$\hat{I}$	peak current
I'	moment of inertia about a parallel axis
I_0	activity, initial (radioactivity)
I_0	extraterrestrial solar radiant intensity
I_{av}	average current
I_c	capacitive current
I_e	radiant intensity [IUPAP]
I_L	inductive current
I_m	maximum current
I_m	maximum peak current
I_{mp}	maximum peak current
I_n	equivalent noise current
I_p	peak current
I_{pk}	peak current
I_R	current through resistance
I_s	saturation current
I_T	total current
I_{xx}	moment of inertia
I_{xy}	product of inertia
j	heat transfer factor
j	imaginary unit
j	imaginary unit [IUPAP]
j	magnetic dipole moment [IUPAP]
j	quantum number, inner
j	right-angle turning operator
j	right-angle turning operator [IEC]
j	square root of minus one
j_i	total angular momentum quantum number [IUPAP]
j_m	mass transfer factor [British]
$\mathbf{j}$	magnetic dipole moment [IUPAP]
$\mathbf{j}$	unit vector along Y-axis
J	action variable
J	direct solar radiant intensity, in or below atmosphere
J	electric current density [IUPAC]
J	electric current density [IUPAP]
J	electric equivalent of heat
J	emissive power, total

DPMA

J		Joule equivalent
J		magnetization [IEC]
J		mechanical equivalent of heat
J		moment of inertia
J		moment of inertia [IEC]
J		moment of inertia [IUPAP]
J		number of equivalents
J		polar moment of inertia
J		radiant intensity
J		rotational quantum number [IUPAP]
J		second zonal harmonic coefficient
J		total angular momentum quantum number [IUPAP]
J		total inner quantum number
J_0		extraterrestrial solar radiant intensity
J_L		load moment of inertia
J_λ		emissive power, monochromatic
J_λ		spectral radiant intensity
J_ν		emissive power, monochromatic
J		electric current density [IUPAP]
J		magnetic polarization [IUPAP]
J		electric current density
k		Boltzmann's constant
k		Boltzmann's constant [British]
k		Boltzmann's constant [IUPAC]
k		Boltzmann's constant [IUPAP]
k		circular wave number [ISO]
k		circular wave number [IUPAP]
k		coefficient
k		coefficient of compressibility
k		compressibility factor
k		coupling coefficient [IEC]
k		coupling factor
k		electromechanical coupling factor
k		factor
k		gas constant, molecular
k		kinetic (subscript)
k		load per unit displacement
k		magnetic susceptibility
k		radius of gyration
k		radius of gyration [British]
k		ratio of specific heats [British]
k		reaction velocity constant

k	restoring force per unit displacement	
k	specific reaction rate	
k	spring constant	
k	thermal conductivity	
k	thermal conductivity [British]	
k	torque per unit twist	
k	torsion constant	
k	velocity constant of reaction [British]	
k	wavelength constant	
k	wave number	
k^{-1}	thermal resistivity	
k_0	Von Karman's constant	
k_{15}	shear coupling factor (piezoelectricity)	
k_{31}	transverse coupling factor (piezoelectricity)	
k_{33}	longitudinal coupling factor (piezoelectricity)	
k_g	grid conductance at zero frequency and constant plate potential	
k_p	planar coupling factor (piezoelectricity)	
k_p	plate conductance at zero frequency and constant grid potential	
k_t	thickness coupling factor (piezoelectricity)	
k_T	thermal diffusion ratio	
k	kinetic (subscript) [IUPAP]	
k	Boltzmann's constant [IUPAC]	
k	unit vector along Z-axis	
K	bulk modulus [British]	
K	bulk modulus [IUPAP]	
K	bulk modulus of elasticity	
K	cathode terminal	
K	circulation [British]	
K	coefficient of eddy transfer, kinematic	
K	compression modulus [IUPAC]	
K	curvature	
K	electric field strength [IEC]	
K	electric intensity [IEC]	
K	equilibrium constant	
K	equilibrium constant [IUPAC]	
K	equilibrium constant [IUPAP]	
K	equilibrium constant of reaction [British]	
K	Kerr constant	
K	luminosity factor	
K	luminous efficiency	

K	magnetostriction constant	
K	rotational quantum number [IUPAP]	
K	stress concentration factor	
K	vortex strength [British]	
K_a	equilibrium constant expressed in terms of activity [British]	
K_a	equilibrium constant in terms of activity	
K_c	equilibrium constant expressed in terms of concentration [British]	
K_h	hydrolysis constant	
K_p	equilibrium constant expressed in terms of pressure [British]	
K_T	thermal diffusion ratio [IUPAP]	
K_λ	luminosity	
K_λ	monochromatic luminous efficiency	
K_λ	visibility factor	
Kn	Knudsen number [British]	
l	Coriolis parameter	
l	direction cosine	
l	distance	
l	free path	
l	heat flow path, length of	
l	latent heat, specific [British]	
l	length	
l	length [British]	
l	length [IEC]	
l	length [ISO]	
l	length [IUPAC]	
l	length [IUPAP]	
l	length of vibrating string, rod, or tube	
l	mean free path [British]	
l	mean free path [IUPAC]	
l	mean free path [IUPAP]	
l	mixing length	
l	neutron lifetime [British]	
l	quantum number, azimuthal or orbital	
l	scale of turbulence [British]	
l	mean free path	
l_0	rest length	
l_i	orbital angular momentum quantum number [IUPAP]	
l_T	Tait free path	

DPMA

L	Avogadro's constant [IUPAP]	
L	Avogadro's number [IUPAC]	
L	distance	
L	heat flow path, length of	
L	inductance	
L	inductance [British]	
L	inductance [IUPAP]	
L	kinetic potential	
L	Lagrangian function	
L	Lagrangian function [IUPAP]	
L	latent heat of phase change	
L	left (subscript)	
L	length	
L	linear energy transfer	
L	liquid (subscript)	
L	Lorentz unit	
L	luminance [British]	
L	luminance [IUPAC]	
L	luminance [IUPAP]	
L	luminous emittance	
L	orbital angular momentum quantum number [IUPAP]	
L	path [IUPAP]	
L	photometric brightness [British]	
L	quantum number, total azimuthal or orbital	
L	self-inductance	
L	self-inductance [British]	
L	self-inductance [IEC]	
L	self-inductance [IUPAC]	
L	self-inductance [IUPAP]	
L_{12}	mutual inductance	
L_{12}	mutual inductance [IUPAC]	
L_{12}	mutual inductance [IUPAP]	
L_c	conversion loss	
L_e	radiance [IUPAP]	
L_M	molecular-scale temperature geometric gradient	
L'_M	molecular-scale temperature geopotential gradient	
L_{mr}	mutual inductance [British]	
L_{mn}	mutual inductance [IEC]	
L_N	loudness level [IUPAP]	
L	liquid (subscript) [British]	
m	direction cosine	

DPMA

m	....	electromagnetic moment [IUPAP] ....
m	....	electron mass [IUPAP] ....
m	....	flare coefficient in a horn ....
m	....	magnetic pole strength ....
m	....	magnification, linear ....
m	....	mass ....
m	....	mass [British] ....
m	....	mass [IEC] ....
m	....	mass [IUPAC] ....
m	....	mass [IUPAP] ....
m	....	mass (subscript) ....
m	....	mass of electron ....
m	....	mass of electron [British] ....
m	....	maximum (subscript) ....
m	....	mean (subscript) ....
m	....	modulation factor ....
m	....	molality ....
m	....	molality [British] ....
m	....	molality [IUPAC] ....
m	....	molality of solution [IUPAP] ....
m	....	molecular mass ....
m	....	molecular mass [IUPAC] ....
m	....	molecular mass [IUPAP] ....
m	....	number of phases ....
m	....	number of phases [IEC] ....
m	....	optical air mass ....
m	....	optical air mass (subscript) ....
m	....	order of spectrum ....
m	....	phase number [IUPAP] ....
m	....	quantum number, magnetic ....
$\dot{m}$	....	flow rate, mass [British] ....
$\dot{m}$	....	mass flow rate ....
$\bar{m}$	....	molecular mass ....
m_0	....	rest mass ....
m_e	....	electron mass ....
m_e	....	electron mass [IUPAP] ....
m_l	....	magnetic quantum number [IUPAP] ....
m_m	....	mass of atom ....
m_m	....	mass of molecule ....
m_n	....	neutron mass [IUPAP] ....
m_p	....	proton mass [IUPAP] ....
m_u	....	unified atomic mass constant [IUPAP] ....

m_μ	meson mass [IUPAP]	
m_π	meson mass [IUPAP]	
m	mean value (subscript) [British]	
max	maximum (subscript)	
min	minimum (subscript)	
m	electromagnetic moment [IUPAP]	
m	magnetic moment	
M	atomic mass [IUPAP]	
M	atomic weight	
M	bending moment	
M	bending moment [British]	
M	inertance	
M	intensity of magnetization [British]	
M	luminous emittance [IUPAP]	
M	Mach number	
M	Mach number [British]	
M	magnetic polarization [British]	
M	magnetic quantum number [IUPAP]	
M	magnetization [IUPAC]	
M	magnetization [IUPAP]	
M	magnetomotive force [British]	
M	molar mass	
M	molar mass [IUPAC]	
M	molecular weight	
M	molecular weight [British]	
M	moment	
M	moment [British]	
M	moment of force	
M	moment of force [IEC]	
M	moment of force [IUPAC]	
M	moment of force [IUPAP]	
M	momentum	
M	momentum (subscript)	
M	mutual inductance	
M	mutual inductance [British]	
M	mutual inductance [IEC]	
M	mutual inductance [IUPAC]	
M	mutual inductance [IUPAP]	
M	nuclear mass [IUPAP]	
M	oscillator figure of merit	
M	quantum number, total magnetic	
M_a	atomic mass [IUPAP]	

M_B	molar mass of substance B [IUPAP]	
M_e	radiant emittance [IUPAP]	
M_N	nuclear mass [IUPAP]	
M/z	equivalent weight	
$\boldsymbol{M}$	magnetization [IUPAP]	
$\boldsymbol{M}$	moment of force [IUPAP]	
$\boldsymbol{M}$	magnetic polarization	
$\boldsymbol{M}$	magnetization	
n	amount of substance [IUPAP]	
n	angular frequency without damping	
n	circular wave number	
n	direction cosine	
n	molecular concentration	
n	molecular density	
n	neutron density [British]	
n	neutrons, number of [British]	
n	normal (subscript)	
n	normal component (subscript)	
n	number	
n	number [British]	
n	number density	
n	number density of molecules [IUPAP]	
n	number in a sample [British]	
n	number of components (Gibbs phase rule)	
n	number of molecules per unit volume	
n	number of moles	
n	number of moles [British]	
n	number of moles [IUPAC]	
n	number of objects	
n	number of observations	
n	number of revolutions per unit time	
n	number of revolutions per unit time [IEC]	
n	order of spectrum	
n	perpendicular (subscript)	
n	polytropic exponent	
n	polytropic index [British]	
n	principal quantum number [IUPAP]	
n	quantum number, principal	
n	refraction index [IUPAC]	
n	refractive index	
n	refractive index [British]	
n	refractive index [IUPAP]	

n		revolutions per unit time
n		rigidity
n		rotational frequency
n		rotational frequency [British]
n		rotational frequency [IEC]
n		rotational frequency [ISO]
n		rotations per unit time
n		sample size
n		shear modulus of elasticity
n		speed of rotation [IEC]
n		total moles
n'		angular frequency of free vibration with damping
n_0		Loschmidt number
n_0		molecular concentration at zero degree centigrade and one atmosphere
n_g		refractive index, group
n_i		principal quantum number [IUPAP]
n		neutron
n		neutron [IUPAP]
n		normal (subscript) [IUPAP]
n		standard value (subscript) [British]
N		Avogadro's number
N		Avogadro's number [British]
N		Avogadro's number [IUPAC]
N		cloud amount
N		dimensionless number
N		factor of safety
N		frequency constant
N		grating lines, total number of
N		load factor [British]
N		neutron number [IUPAP]
N		neutrons in nucleus, number of [British]
N		normalization factor
N		number
N		number of atoms or nuclei at specified time
N		number of atoms per volume of a particular compound [British]
N		number of atoms per volume of a particular element [British]
N		number of conductors
N		number of molecules
N		number of molecules [British]

N	number of molecules [IUPAC]	
N	number of molecules [IUPAP]	
N	number of molecules, total	
N	number of neutrons	
N	number of turns	
N	number of turns [British]	
N	number of turns [IUPAP]	
N	number of turns in a winding [IEC]	
N	radiance	
N	steradiance	
N	total number of turns	
N'	number of atoms per volume of a mixture [British]	
N_0	Avogadro's number	
N_0	Avogadro's number [IUPAC]	
N_0	initial number of atoms	
N_0	initial number of nuclei	
N_A	Avogadro's constant [IUPAP]	
N_P	Prandtl number	
N_λ	spectral radiance	
Nu	Nusselt number	
Nu	Nusselt number [British]	
o	initial value (subscript)	
o	output (subscript)	
o	outside (subscript)	
o	reference conditions (subscript)	
o	standard value (subscript)	
O	origin	
p	angular frequency of impressed force	
p	atmospheric pressure	
p	constant pressure (subscript)	
p	constant pressure (subscript) [British]	
p	distance between a principal plane of a lens system and the appropriate principal plane of a unit	
p	electric dipole moment [IUPAP]	
p	generalized momentum	
p	generalized momentum [IUPAP]	
p	impact parameter	
p	isobaric processes (subscript)	
p	magnetic poles, number of pairs of	
p	magnetic pole strength	
p	momentum [British]	
p	momentum [IUPAP]	

p	osmotic pressure	
p	partial potential coefficient	
p	partial pressure	
p	plate (subscript)	
p	potential (subscript)	
p	power, instantaneous	
p	pressure	
p	pressure [British]	
p	pressure [IEC]	
p	pressure [IUPAC]	
p	pressure [IUPAP]	
p	propagation coefficient [IEC]	
p	resonance escape probability [British]	
p	semilatus rectum of conic section	
p	sound pressure, root-mean-square	
p	specific pressure	
p	statistical weight [IUPAC]	
p	vapor pressure	
p	vapor pressure [British]	
p^*	vapor pressure	
p_0	total pressure [British]	
p_a	atmospheric pressure	
p_a	sound pressure, average	
p_a	atmospheric pressure [British]	
p_{at}	atmospheric pressure [British]	
p_c	critical pressure	
p_i	sound pressure, instantaneous	
p_m	sound pressure, maximum	
p_p	sound pressure, peak	
p_r	reduced pressure	
p_t	total pressure [British]	
p	proton	
p	proton [IUPAP]	
p	electric dipole moment [IUPAP]	
p	molecular momentum vector [IUPAP]	
p	momentum [IUPAP]	
P	active power	
P	active power [IEC]	
P	amplitude of simple harmonic pressure	
P	concentrated force	
P	concentrated load	
P	dielectric polarization [IUPAC]	

P		dielectric polarization [IUPAP]
P		electric polarization [British]
P		perimeter
P		permeance [IEC]
P		porosity
P		power
P		power [British]
P		power [IEC]
P		power [IUPAC]
P		power [IUPAP]
P		pressure [IUPAC]
P		probability
P		proton number [IUPAP]
P		radiant flux
P		radiant flux [IUPAP]
P		radiant power
P		sound energy flux [IUPAP]
P		total load
P		total pressure
P_0		total pressure [British]
P_a		apparent power
P_A		acoustical power
P_{at}		atmospheric pressure [British]
P_E		electrical power
P_i		input power
P_M		mechanical power
P_o		output power
P_o		pressure, static
P_p		peak power
P_p		plate power
P_p		primary power
P_q		reactive power
P_q		reactive power [IEC]
P_R		rotational power
P_s		apparent power [IEC]
P_s		secondary power
P_s		strength of surface double layer
P_t		total pressure [British]
P_λ		spectral radiant power
Pe		Peclet number [British]
Pr		Prandtl number
Pr		Prandtl number [British]

P	pseudoscalar coupling [IUPAP]	
P	dielectric polarization [IUPAP]	
P	electric polarization	
P_s	strength of double layer	
P_s	surface polarization	
q	charge, instantaneous	
q	discharge rate	
q	electric charge	
q	generalized coordinate	
q	generalized coordinate [IUPAP]	
q	heat [IUPAC]	
q	heat (subscript)	
q	heat entering system	
q	heat flow rate	
q	heat flux intensity [British]	
q	load per unit distance	
q	quantity of heat [British]	
q	shear stress [British]	
q	specific humidity	
q_s	sensible heat [British]	
Q	aggregate quantity of heat [British]	
Q	concentrated force	
Q	concentrated load	
Q	disintegration energy [IUPAP]	
Q	electric charge	
Q	electric charge [British]	
Q	electric quadrupole moment	
Q	energy balance in a nuclear reaction	
Q	energy storage factor	
Q	figure of merit of inductance	
Q	flow rate, volume [British]	
Q	heat [IUPAC]	
Q	luminous energy	
Q	moment of area	
Q	partition function [British]	
Q	partition function [IUPAC]	
Q	partition function [IUPAP]	
Q	power, reactive [British]	
Q	Q-factor [British]	
Q	quadrupole moment [IUPAP]	
Q	quality factor of reactor	
Q	quantity	

Q	quantity of electric charge [IEC]	
Q	quantity of electricity [IEC]	
Q	quantity of electricity [IUPAC]	
Q	quantity of electricity [IUPAP]	
Q	quantity of heat	
Q	quantity of heat [British]	
Q	quantity of heat [IUPAP]	
Q	quantity of light	
Q	quantity of light [IUPAC]	
Q	quantity of light [IUPAP]	
Q	ratio of reactance to resistance	
Q	reaction energy [IUPAP]	
Q	reactive power [IEC]	
Q	shear force [British]	
Q	strength of source	
Q	thermoelectric power	
Q	total solar and sky radiant intensity	
Q	volumetric flow rate	
$\dot{Q}$	heat flow rate	
$\dot{Q}$	flow rate, heat [British]	
$\dot{Q}$	rate of heat transfer [British]	
Q_e	quantity of radiant energy [IUPAP]	
Q_s	sensible heat [British]	
r	angle of reflection	
r	angle of refraction	
r	capacitance ratio	
r	gas constant, specific	
r	mole ratio of solution [IUPAP]	
r	nuclear radius	
r	nuclear radius [British]	
r	radial (subscript)	
r	radial distance	
r	**radius**	
r	radius [British]	
r	radius [IEC]	
r	radius [ISO]	
r	radius [IUPAC]	
r	radius [IUPAP]	
r	radius (subscript)	
r	radius of curvature	
r	radius of gyration	
r	radius of gyration [British]	

DPMA

r	reduced properties (subscript)	
r	refractivity [IUPAC]	
r	relative humidity	
r	resistance	
r	resistivity [British]	
r	root (subscript)	
r	rotational (subscript)	
r	specific resistance [British]	
r_A	acoustical resistance	
r_{Al}	specific acoustical resistance	
r_e	electron radius [IUPAP]	
r_E	electrical resistance	
r_M	mechanical resistance	
r_P	plate resistance	
r_p	pressure ratio [British]	
r_R	rotational resistance	
r_V	volume ratio [British]	
$\mathbf{r}$	radiation (subscript) [British]	
$\mathbf{r}$	relative (subscript) [IUPAP]	
$\mathbf{r}$	molecular position vector [IUPAP]	
$\mathbf{r}$	position vector	
$\mathbf{r}$	radius vector	
R	gas constant	
R	gas constant [IUPAC]	
R	gas constant, characteristic [British]	
R	gas constant per mole [IUPAP]	
R	gas constant, specific	
R	gas constant, specific [British]	
R	gas constant, universal	
R	linear range [IUPAP]	
R	modulus of rupture	
R	modulus of rupture [British]	
R	nuclear radius [British]	
R	nuclear radius [IUPAP]	
R	quantum number, rotational	
R	radiation (subscript)	
R	radius of Earth	
R	range	
R	ratio	
R	reluctance [British]	
R	reluctance [IEC]	
R	resistance	

R	resistance [British]	
R	resistance [IEC]	
R	resistance [IUPAC]	
R	resistance [IUPAP]	
R	responsivity	
R	Reynolds number	
R	Reynolds number [British]	
R	right (subscript)	
R	Rydberg constant	
R	Rydberg's constant [British]	
R	thermal resistance	
R	turbulence-correlation coefficient	
R^*	gas constant, universal	
R_∞	Rydberg constant [IUPAP]	
R_∞	Rydberg constant for infinite mass	
R_0	gas constant, molar [British]	
R_0	gas constant, universal [British]	
R_A	acoustical resistance	
R_A	acoustical response	
R_E	electrical resistance	
R_E	electrical response	
R_g	grid resistance	
R_k	cathode resistance	
R_l	linear range [IUPAP]	
R_L	load resistance	
R_m	reluctance [IEC]	
R_M	gas constant for gas of molecular weight M [British]	
R_M	mechanical resistance	
R_M	mechanical response	
R_o	gas constant, molar	
R_R	rotational resistance	
R_R	rotational response	
R_s	squareness ratio	
R_θ	directivity ratio	
Ra	Rayleigh number [British]	
Re	Reynolds number	
Re	Reynolds number [British]	
Ri	Richardson number	
R	gas constant [IUPAC]	
R	gas constant, molar [British]	
R	gas constant, universal [British]	

Symbol	Description
$\mathcal{R}$	gas constant, characteristic [British]
$\mathcal{R}$	gas constant, specific [British]
$\mathcal{R}$	radiance
$\mathcal{R}$	radiancy
$\mathcal{R}$	radiant flux density
$\mathcal{R}$	reluctance
$\mathcal{R}$	reluctance [IEC]
s	adiabatic (subscript)
s	angular momentum, intrinsic
s	arc length [IUPAC]
s	arc length distance
s	arc of curve [IEC]
s	condensation
s	constant entropy (subscript) [British]
s	curved path [IEC]
s	displacement
s	distance
s	distance between object and corresponding principal plane
s	dynamic sensitivity of a phototube
s	elastic compliance
s	entropy
s	entropy per atom
s	entropy per mole
s	entropy per molecule
s	entropy per unit mass
s	grid, screen (subscript)
s	length
s	length of arc
s	length of path
s	length of path [ISO]
s	linear distance
s	normal stress
s	object distance
s	path [IUPAC]
s	path [IUPAP]
s	pseudo-adiabatic processes (with loss of condensate) (subscript)
s	quantum number, spin
s	saturation (subscript)
s	scattering coefficient
s	slip (electrical machinery)

s	....	slip [IEC]
s	....	source (subscript)
s	....	specific entropy
s	....	specific entropy [British]
s	....	spin
s	....	stability (subscript)
s	....	standard deviation of sample
s	....	stationary (subscript)
s	....	stiffness
s	....	streamline (subscript)
s	....	sublimation process (subscript)
s	....	surface (subscript)
s	....	surface area per unit mass
s	....	turbidity
s'	....	distance beween image and corresponding principal plane
s'	....	image distance
s_i	....	spin quantum number [IUPAP]
s_m	....	entropy per atom
s_m	....	entropy per molecule
s_p	....	transconductance at zero frequency and constant plate potential
s_s	....	shear stress
s, n, z	....	natural coordinates
sp	....	specific (subscript)
su	....	suppressor grid (subscript)
s	....	sublimation (subscript) [British]
s	....	surface in contact with substance (subscript) [British]
s	....	linear displacement
S	....	action
S	....	apparent power [British]
S	....	apparent power [IEC]
S	....	area
S	....	area [IEC]
S	....	area [ISO]
S	....	area [IUPAC]
S	....	area [IUPAP]
S	....	area of a diaphragm
S	....	area of a radiator
S	....	area of a tube
S	....	area of cross-section of walls

S	constant entropy (subscript)	
S	constant strain (superscript)	
S	elastance	
S	entropy	
S	entropy [British]	
S	entropy [IUPAC]	
S	entropy [IUPAP]	
S	entropy per mole	
S	linear stopping power [IUPAP]	
S	Poynting vector	
S	Poynting vector [British]	
S	Poynting vector [IUPAP]	
S	projected area	
S	quantum number, total spin	
S	reciprocal capacitance	
S	reluctance [British]	
S	salinity	
S	saturation	
S	sedimentation coefficient	
S	self-elastance	
S	spin quantum number [IUPAP]	
S	spin, total	
S	static sensitivity of a phototube	
S	Sutherland's constant	
S	total entropy	
S	total entropy [British]	
S_a	atomic stopping power [IUPAP]	
S_l	linear stopping power [IUPAP]	
S_M	entropy per mole	
$\mathbf{S}$	scalar coupling [IUPAP]	
$\mathbf{s}$	solid (subscript) [British]	
$\mathbf{S}$	Poynting vector [IUPAP]	
t	customary temperature [British]	
t	empirical temperature [British]	
t	length of prism base	
t	tangential component (subscript)	
t	temperature	
t	temperature [IEC]	
t	temperature [IUPAC]	
t	temperature [IUPAP]	
t	temperature, ordinary	
t	time	

t	time [British]	
t	time [IEC]	
t	time [ISO]	
t	time [IUPAC]	
t	time [IUPAP]	
t	total (subscript)	
t	trajectory (subscript)	
t	transition between polymorphic forms (subscript)	
t	transport number [British]	
t	transport number [IUPAC]	
t_0	temperature of ice point, ordinary	
t_c	critical temperature	
t_m	mean-life of radioisotope [British]	
t	transition (subscript) [British]	
t	triton	
t	triton [IUPAP]	
T	absolute temperature	
T	absolute temperature [British]	
T	absolute temperature [IEC]	
T	absolute temperature [IUPAC]	
T	constant stress (superscript)	
T	constant temperature (subscript)	
T	constant temperature (subscript) [British]	
T	force in membrane	
T	force in string	
T	half-life (radioactivity)	
T	half-life of radioisotope [British]	
T	isotopic spin [British]	
T	Kelvin temperature	
T	kinetic energy	
T	kinetic energy [British]	
T	kinetic energy [IUPAP]	
T	oscillation period	
T	period	
T	period [British]	
T	period [IUPAP]	
T	periodic time [ISO]	
T	periodic time [IUPAC]	
T	reverberation time	
T	temperature	
T	thermodynamic temperature	
T	thermodynamic temperature [IUPAP]	
T	thrust (subscript)	

DPMA

T	time constant [IEC]	
T	time interval [IUPAC]	
T	time of one cycle [IEC]	
T	torque	
T	torque [British]	
T	torque [IEC]	
T	transmittance [IUPAC]	
T	transport number [British]	
T	transport number [IUPAC]	
T^*	absolute virtual temperature	
T_0	absolute temperature of ice point	
$T_{½}$	half-life [IUPAP]	
$T_{½}$	half-life of radioisotope [British]	
T_c	critical temperature	
T_c	Curie temperature	
T_t	total (absolute) temperature [British]	
T_v	absolute virtual temperature	
T	tensor coupling [IUPAP]	
u	angle of slope in object space	
u	component of particle velocity in x direction	
u	density of radiant energy	
u	energy of gas, specific internal [British]	
u	energy per atom, internal	
u	energy per atom, intrinsic	
u	energy per molecule, internal	
u	energy per molecule, intrinsic	
u	energy per mole, internal	
u	energy per mole, intrinsic	
u	energy per unit mass, internal	
u	energy per unit mass, intrinsic	
u	lethargy [British]	
u	linear velocity [British]	
u	particle velocity, root-mean-square	
u	reaction velocity	
u	specific internal energy	
u	specific internal energy [British]	
u	speed	
u	speed at specified time	
u	speed, linear	
u	speed, particle	
u	time-independent wave function	
u	transverse component of velocity	

u		velocity
u		velocity [ISO]
u		velocity [IUPAC]
u		velocity [IUPAP]
u		velocity, ionic
u		x component of velocity
$\bar{u}$		average speed
$\bar{u}$		average velocity [IUPAP]
$\tilde{u}$		speed, root-mean-square
$\tilde{u}$		velocity, root-mean-square
$\hat{u}$		most probable speed [IUPAP]
$\hat{u}$		speed, most probable
u'		angle of slope in image space
u'		x component of perturbation velocity
u^*		friction velocity
u_a		particle velocity, average
u_{av}		average speed
u_i		particle velocity, instantaneous
u_m		energy per atom, internal
u_m		energy per atom, intrinsic
u_m		energy per molecule, internal
u_m		energy per molecule, intrinsic
u_m		particle velocity, maximum
u_o		initial speed
u_p		particle velocity, peak
u_t		speed at specified time
u_w		wave velocity
u_φ		phase velocity
$\mathbf{u}$		molecular velocity vector [IUPAP]
$\mathbf{u}_0$		average velocity [IUPAP]
u		linear velocity
u		velocity
u		velocity at time t
u		velocity, group
u		velocity, particle
$\bar{u}$		velocity, average
$\hat{u}$		velocity, most probable
u_{av}		velocity, average
u_g		velocity, group
u_o		initial velocity
u_t		velocity at time t
U		energy [IEC]

U	energy [IUPAC]	
U	energy of gas, internal [British]	
U	energy per mole	
U	energy per mole, internal	
U	energy per mole, intrinsic	
U	energy, total	
U	internal energy	
U	internal energy [British]	
U	internal energy [IUPAP]	
U	internal energy, total [British]	
U	internal energy, total value	
U	intrinsic energy	
U	intrinsic energy, total value	
U	over-all heat transfer coefficient	
U	over-all heat transfer coefficient [British]	
U	potential difference [IEC]	
U	potential energy	
U	radiant energy	
U	relative humidity	
U	strain energy [British]	
U	volume current	
U	volume velocity	
U_M	energy per mole, internal	
U_M	energy per mole, intrinsic	
U_λ	spectral radiant energy	
U_τ	friction velocity [British]	
U_τ	shear velocity [British]	
v	component of particle velocity in y direction	
v	constant volume (subscript)	
v	constant volume (subscript) [British]	
v	difference of electric potential	
v	linear velocity [British]	
v	linear velocity [IEC]	
v	phase velocity of electromagnetic waves [British]	
v	potential difference, instantaneous	
v	potential difference, instantaneous [British]	
v	quantum number, vibrational	
v	radial component of velocity	
v	radial velocity	
v	specific volume	
v	specific volume [British]	
v	specific volume [IUPAC]	

v	speed	
v	speed at specified time	
v	speed, linear	
v	speed of propagation	
v	speed, particle	
v	vaporization processes (subscript)	
v	velocity	
v	velocity [British]	
v	velocity [ISO]	
v	velocity [IUPAC]	
v	velocity [IUPAP]	
v	velocity of sound or other waves	
v	vertical (subscript)	
v	vibrational quantum number [IUPAP]	
v	virtual quantities (subscript)	
v	volume [British]	
v	volume [ISO]	
v	volume [IUPAC]	
v	volume [IUPAP]	
v	volume per atom	
v	volume per mole	
v	volume per molecule	
v	volume per unit mass	
v	y component of velocity	
$\bar{v}$	average speed	
$\tilde{v}$	speed, root-mean-square	
$\tilde{v}$	velocity, root-mean-square	
$\hat{v}$	speed, most probable	
v'	y component of perturbation velocity	
v^*	friction velocity [British]	
v^*	shear velocity [British]	
v_{av}	average speed	
v_m	volume per atom or molecule	
v_o	initial speed	
v_t	speed at specified time	
v	linear velocity	
v	velocity	
v	velocity at time t	
v	velocity, group	
v	velocity, particle	
v̄	velocity, average	
v̂	velocity, most probable	

v_{av}	velocity, average	
v_g	velocity, group	
v_o	initial velocity	
v_t	velocity at time t	
V	atomic volume	
V	configuration-space volume	
V	effective potential difference	
V	electric potential	
V	electric potential [British]	
V	electric potential [IUPAC]	
V	electric potential [IUPAP]	
V	electromotive force	
V	electromotive force [British]	
V	molecular volume	
V	potential [IEC]	
V	potential difference	
V	potential difference [IEC]	
V	potential difference, root-mean-square	
V	potential difference, steady direct-current	
V	potential energy	
V	potential energy [British]	
V	potential energy [IUPAP]	
V	potential, inner	
V	shearing force in beam section	
V	specific magnetic rotation	
V	time-dependent variable in Hamilton-Jacobi equation	
V	Verdet constant	
V	vertical reaction	
V	voltage	
V	voltage, maximum	
V	volume	
V	volume [British]	
V	volume [IEC]	
V	volume [ISO]	
V	volume [IUPAC]	
V	volume [IUPAP]	
V	volume of a cavity or room	
V	volume per mole	
$\dot{V}$	flow rate, volume [British]	
$\bar{V}$	potential difference, average	
$\bar{V}$	quiescent potential difference	

$\hat{V}$	peak potential difference	
V_c	critical volume	
V_e	excitation potential difference	
V_i	inner potential	
V_i	ionization potential	
V_{ij}	interaction energy between molecules i and j [IUPAP]	
V_m	maximum potential difference	
V_m	specific volume [British]	
V_M	volume per mole	
V_{mp}	maximum peak potential difference	
V_p	peak potential difference	
V_{pk}	peak potential difference	
V_s	Seebeck potential difference	
V_t	Thomson potential difference	
V_v	potential difference, contact or Volta	
V_π	Peltier potential difference	
$\hat{V}_m$	maximum peak potential difference	
V	vector coupling [IUPAP]	
w	component of particle velocity in z direction	
w	linear velocity [British]	
w	mass fraction	
w	mixing ratio	
w	phase velocity	
w	range [British]	
w	velocity [ISO]	
w	velocity [IUPAC]	
w	velocity potential, complex [British]	
w	wall (subscript)	
w	water vapor (subscript)	
w	water-vapor content	
w	wave velocity	
w	weight	
w	weight density [British]	
w	weight flow per unit time	
w	wet-bulb temperature (subscript)	
w	width	
w	work	
w	work [British]	
w	work [IUPAC]	
w	work function, net	
w	z component of velocity	
w'	z component of perturbation velocity	

w_B		mass fraction of substance B [IUPAP]
w_g		work function, gross
w		wall (subscript) [British]
W		average energy expended in a gas per ion pair formed
W		electrical energy [British]
W		energy [British]
W		energy [IEC]
W		energy required to form one ion pair [British]
W		force due to gravity [IUPAC]
W		irradiance
W		precipitable water
W		radiancy
W		radiant emittance
W		radiant flux density
W		time-dependent variable in Hamilton-Jacobi equation
W		total load
W		weight
W		weight [British]
W		weight [IUPAC]
W		weight [IUPAP]
W		work
W		work [IEC]
W		work [IUPAP]
W_λ		spectral radiant emittance
x		coordinate along X-axis
x		distance between object and principal focus of object space
x		mole fraction
x		mole fraction [British]
x		mole fraction [IUPAC]
x		mole fraction basis (subscript)
$\bar{x}$		average value of a quantity x
x'		distance between image and principal focus of image space
x_A		acoustical reactance
x_B		mole fraction of substance B [IUPAP]
x_E		electrical reactance
x_M		mechanical reactance
x_R		rotational reactance
x, y, z		orthogonal coordinates, Cartesian

x, y, z	x, y, z coordinates (subscript)	
X	adiabatic factor	
X	horizontal axis	
X	mole fraction [IUPAC]	
X	observed value	
X	reactance	
X	reactance [British]	
X	reactance [IEC]	
X	reactance [IUPAC]	
X	reactance [IUPAP]	
X	solid (subscript)	
X	volume displacement	
$[X]$	molar concentration of substance X [British]	
$\{X\}$	relative activity of substance X [British]	
X_A	acoustical reactance	
X_B	mole fraction of substance B [IUPAP]	
X_C	capacitive reactance	
X_E	electrical reactance	
X_L	inductive reactance	
X_M	mechanical reactance	
X_R	rotational reactance	
X	crystalline (subscript) [British]	
y	altitude	
y	coordinate along Y-axis	
y	depth	
y	height	
y	mole fraction [IUPAC]	
y	mole fraction, gas phase	
y	object height	
y	object length	
y	supercompressibility factor	
y	thickness	
y'	height of image	
y'	length of image	
y_A	acoustical admittance	
y_E	electrical admittance	
y_g	grid admittance with plate load	
y_M	mechanical admittance	
y_p	output admittance	
y_p	plate admittance	
y_{pg}	grid-plate transadmittance	
y_R	rotational admittance	

Y	admittance	
Y	admittance [British]	
Y	admittance [IEC]	
Y	admittance [IUPAP]	
Y	reciprocal impedance	
Y	vertical axis	
Y	Young's modulus of elasticity	
Y_A	acoustical admittance	
Y_E	electrical admittance	
Y_M	mechanical admittance	
Y_R	rotational admittance	
z	altitude	
z	charge number of an ion [IUPAC]	
z	charge number of ion [IUPAP]	
z	coordinate along Z-axis	
z	electrochemical equivalent	
z	head, potential [British]	
z	impedance	
z	self-impedance	
z	valence	
z	valency of an ion [British]	
$\bar{z}$	complex conjugate of z [IUPAP]	
$|z|$	modulus of z [IUPAP]	
z'	real part of z [IUPAP]	
z''	imaginary part of z [IUPAP]	
z^*	complex conjugate of z [IUPAP]	
z^*	conjugate of z [IUPAP]	
z_0	roughness parameter	
z_A	acoustical impedance, complex	
z_B	activity of substance B [IUPAP]	
z_E	electrical impedance, complex	
z_M	mechanical impedance, complex	
z_p	pressure altitude	
z_R	rotational impedance, complex	
Z	atomic number	
Z	atomic number [British]	
Z	atomic number [IUPAP]	
Z	collision frequency, molecular	
Z	impedance	
Z	impedance [British]	
Z	impedance [IEC]	
Z	impedance [IUPAC]	

DPMA

Z		impedance [IUPAP]
Z		modulus of section
Z		modulus of section [British]
Z		nuclear charge
Z		partition function [IUPAP]
Z		proton number [IUPAP]
Z		radius of circle of least confusion
Z		radius of image of a point
Z		self-impedance
Z		total vorticity
Z_0		characteristic impedance
Z_0		surge impedance
Z_A		acoustical impedance, complex
Z_E		electrical impedance, complex
Z_i		input impedance
Z_L		load impedance
Z_m		mutual impedance
Z_M		mechanical impedance, complex
Z_o		output impedance
Z_R		rotational impedance, complex
Z_T		total impedance
α		absorptance
α		absorption coefficient of absorbing material
α		absorption factor
α		absorption factor [British]
α		absorption factor [IUPAC]
α		absorption factor [IUPAP]
α		absorptivity
α		acoustic absorption factor [IUPAP]
α		alpha particle
α		alpha particle [IUPAP]
α		angle
α		angle of optical rotation
α		angle of optical rotation [IUPAC]
α		angular acceleration
α		angular acceleration [British]
α		angular acceleration [ISO]
α		angular acceleration [IUPAP]
α		angular resolving power of telescope
α		attenuation coefficient [British]
α		attenuation coefficient [IEC]
α		attenuation coefficient [ISO]
α		attenuation constant

α	coefficient of absorption	
α	coefficient of heat transfer [British]	
α	coefficient of linear expansion	
α	coefficient of recombination	
α	degree of dissociation	
α	degree of dissociation [British]	
α	degree of dissociation [IUPAC]	
α	degree of electrolytic dissociation [IUPAC]	
α	degree of ionization	
α	degree of reaction	
α	degree of reaction [IUPAC]	
α	electric polarizability of a molecule [IUPAC]	
α	end correction for a tube	
α	fine structure constant	
α	fine structure constant [IUPAP]	
α	fraction of incident radiant power which is absorbed [IUPAC]	
α	half-angle subtended at point object by objective of microscope	
α	internal conversion coefficient [IUPAP]	
α	linear coefficient of thermal expansion [British]	
α	linear current density [IEC]	
α	linear expansion coefficient [IUPAP]	
α	linear expansivity	
α	plane angle [IUPAP]	
α	polarizability [IUPAP]	
α	radiant absorption	
α	rotation, specific	
α	Sommerfeld constant	
α	speed, most probable	
α	temperature coefficient [IEC]	
α	temperature coefficient of resistance	
α	thermal diffusivity	
α	thermal diffusivity [British]	
α	velocity, most probable	
α_a	acoustic absorption factor [IUPAP]	
α_T	thermal diffusion factor [IUPAP]	
$\dot{\alpha}$	angular acceleration	
β	Bohr magneton [British]	
β	Bohr magneton [IUPAP]	
β	coefficient of thermal expansion	
β	latitudinal rate of variation of Coriolis parameter	
β	luminance factor [British]	

β	phase-change coefficient [British]	
β	phase coefficient [IEC]	
β	phase coefficient [ISO]	
β	phase constant	
β	plane angle [IUPAP]	
β	ratio of plasma kinetic pressure to magnetic pressure	
β	ratio of speed to speed of light	
β	subsonic compressibility factor [British]	
β	supersonic compressibility factor [British]	
β	thermal coefficient of volumetric expansion	
β	volume expansivity	
γ	activity coefficient [British]	
γ	activity coefficient [IUPAC]	
γ	activity coefficient, molal basis	
γ	activity coefficient, stoichiometric	
γ	activity coefficient, stoichiometric [British]	
γ	actual lapse rate of temperature	
γ	adiabatic exponent	
γ	angular magnification	
γ	conductivity	
γ	conductivity [British]	
γ	conductivity [IEC]	
γ	conductivity [IUPAP]	
γ	contrast of photographic emulsion	
γ	cubic coefficient of thermal expansion [British]	
γ	cubic expansion coefficient [IUPAP]	
γ	electric polarizability of a molecule [IUPAC]	
γ	gyromagnetic ratio [IUPAP]	
γ	photon [IUPAP]	
γ	plane angle [IUPAP]	
γ	polarizability [IUPAP]	
γ	propagation coefficient [British]	
γ	propagation coefficient [IEC]	
γ	propagation coefficient [ISO]	
γ	propagation constant	
γ	ratio of specific heat capacity at constant pressure to specific heat capacity at constant volume [British]	
γ	ratio of specific heats	
γ	ratio of specific heats [British]	
γ	ratio of specific heats [IUPAC]	

DPMA

γ	ratio of specific heats [IUPAP]	
γ	shear strain	
γ	shear strain [British]	
γ	specific conductance [British]	
γ	specific weight	
γ	specific weight [IEC]	
γ	surface tension	
γ	surface tension [British]	
γ	surface tension [IUPAC]	
γ	surface tension [IUPAP]	
γ	weightivity	
γ	weight per unit volume	
γ_B	activity coefficient of substance B [IUPAP]	
Γ	adiabatic lapse rate of temperature	
Γ	circulation	
Γ	circulation [British]	
Γ	gamma function	
Γ	gamma function [British]	
Γ	level width [British]	
Γ	level width [IUPAP]	
Γ	motional capacitance constant	
Γ	reciprocal inductance	
Γ	specific gamma-ray constant	
Γ	strength of single vortex	
Γ	surface concentration [British]	
Γ	surface concentration [IUPAC]	
Γ	vortex strength [British]	
Γ_0	shunt capacitance constant	
Γ_e	electric constant	
Γ_m	magnetic constant	
δ	boundary-layer thickness	
δ	boundary-layer thickness [British]	
δ	damping coefficient	
δ	damping coefficient [IEC]	
δ	damping coefficient [ISO]	
δ	deflection	
δ	deflection [British]	
δ	deflection of beam	
δ	deflection of beam, maximum	
δ	density [IEC]	
δ	departure from normal (subscript)	
δ	deviation	

δ	....	deviation angle
δ	....	dielectric loss angle [British]
δ	....	difference in phase
δ	....	dip angle
δ	....	displacement
δ	....	dissipation factor [IUPAP]
δ	....	frequency deviation
δ	....	galvanometer deflection
δ	....	increment, finite
δ	....	loss angle [IUPAP]
δ	....	piezoelectric strain constant
δ	....	pressure ratio
δ	....	sag of beam
δ	....	thermal expansion, total
δ	....	thickness [ISO]
δ	....	total elongation
δ^*	....	boundary-layer displacement thickness, in two-dimensional and axi-symmetric flow [British]
δ^*	....	displacement thickness
δ_1	....	boundary-layer displacement thickness, in two-dimensional and axi-symmetric flow [British]
δ_2	....	boundary-layer momentum thickness, in two-dimensional and axi-symmetric flow [British]
δ_3	....	boundary-layer energy thickness, in two-dimensional and axi-symmetric flow [British]
δ_m	....	angle of minimum deviation
δx	....	variation of x [IUPAP]
Δ	....	difference in optical path
Δ	....	dilatation
Δ	....	distance between adjacent principal foci of two lens units
Δ	....	increment [IEC]
Δ	....	increment, finite
Δ	....	mass excess [IUPAP]
Δ	....	optical length
Δ	....	optical path difference
Δ	....	optical tube length
Δh	....	heat of reaction or phase change per atom or molecule
Δh	....	heat of reaction or phase change per mole
Δh	....	heat of reaction or phase change per unit mass

Δh_m		heat of reaction or phase change per atom or molecule
ΔH		heat of reaction or phase change per mole
ΔH		heat of reaction or phase change, total value
ΔH_M		heat of reaction or phase change per mole
Δt		temperature difference
Δx		finite increase of x [IUPAP]
ε		base of natural logarithms [IEC]
ε		permittivity [IEC]
ε_0		permittivity of free space [IEC]
ϵ		average molecular kinetic energy
ϵ		base of Naperian logarithms
ϵ		base of natural logarithms [IEC]
ϵ		dielectric coefficient
ϵ		dielectric coefficient, in centimeter-gram-second units
ϵ		dielectric constant [British]
ϵ		dielectric constant [IUPAC]
ϵ		direct strain [British]
ϵ		emissivity [British]
ϵ		emittance
ϵ		epoch angle
ϵ		extinction coefficient
ϵ		horizon distance
ϵ		molar absorptivity [IUPAC]
ϵ		molar decadic absorption [IUPAC]
ϵ		molar extinction coefficient [IUPAC]
ϵ		molecular attraction energy [IUPAP]
ϵ		molecular extinction coefficient [British]
ϵ		normal strain
ϵ		permittivity
ϵ		permittivity [British]
ϵ		permittivity [IEC]
ϵ		permittivity [IUPAC]
ϵ		permittivity [IUPAP]
ϵ		permittivity of medium
ϵ		piezoelectric stress constant
ϵ		radiant emissivity
ϵ		relative dielectric coefficient
ϵ		self energy
ϵ		specific inductive capacity
ϵ		total emissivity [British]

ϵ	turbulence exchange coefficient	
ϵ'	emissivity	
ϵ_0	permittivity of free space	
ϵ_0	permittivity of free space [British]	
ϵ_0	permittivity of free space [IEC]	
ϵ_0	permittivity of vacuum [IUPAP]	
ϵ_r	dielectric coefficient, in meter-kilogram-second units	
ϵ_r	relative dielectric coefficient	
ϵ_r	relative permittivity	
ϵ_r	specific inductive capacity	
ϵ_r	relative permittivity [IUPAP]	
ϵ_t'	total value of emissivity	
ζ	component of particle displacement in the z direction	
ζ	displacement of sound-bearing particle	
ζ	electrokinetic potential [British]	
ζ	electrokinetic potential [IUPAC]	
ζ	vertical component of relative vorticity	
ζ_a	vertical component of absolute vorticity	
η	coefficient of viscosity	
η	component of particle displacement in the y direction	
η	dissipative viscosity	
η	efficiency	
η	efficiency [British]	
η	efficiency [IEC]	
η	efficiency [IUPAP]	
η	electric susceptibility	
η	frictional viscosity	
η	nadir angle	
η	north-south component of horizontal relative vorticity	
η	overpotential [British]	
η	overpotential [IUPAC]	
η	viscosity [IUPAC]	
η	viscosity [IUPAP]	
η^*	object nadir angle	
η_0	minimum nadir angle	
η_1	root of a Bessel equation	
η_r	relative viscosity	
η_v	voltage efficiency	

θ		angle, contact
θ		angle, glancing
θ		angle of diffraction
θ		angle of twist [British]
θ		angular displacement
θ		angular distance
θ		boundary-layer momentum thickness, in two-dimensional and axi-symmetric flow [British]
θ		colatitude
θ		contact angle [IUPAC]
θ		customary temperature [British]
θ		empirical temperature [British]
θ		isothermal processes (subscript)
θ		momentum thickness of boundary layer
θ		normal angle
θ		optical rotation, angle of
θ		phase angle
θ		plane angle
θ		plane angle [IUPAP]
θ		scattering angle [IUPAP]
θ		slope [British]
θ		temperature [IEC]
θ		temperature [IUPAC]
θ		temperature, ordinary
θ		Weiss constant
θ_D		Debye characteristic temperature
ϑ		plane angle [IUPAP]
ϑ		scattering angle [IUPAP]
ϑ		temperature [IEC]
ϑ		temperature [IUPAP]
Θ		absolute temperature
Θ		absolute temperature [IEC]
Θ		characteristic temperature [IUPAC]
Θ		characteristic temperature [IUPAP]
Θ		Debye characteristic temperature
Θ		potential temperature
Θ		thermodynamic temperature [IUPAP]
Θ_D		Debye temperature [IUPAP]
Θ_E		Einstein temperature [IUPAP]
Θ_r		rotational temperature
Θ_r		rotational temperature [IUPAP]

Θ_v	vibrational temperature	
Θ_v	vibrational temperature [IUPAP]	
ι	inclination of orbit	
κ	absorption coefficient [IUPAC]	
κ	absorption index	
κ	compressibility [British]	
κ	compressibility [IUPAC]	
κ	compressibility [IUPAP]	
κ	compressibilty coefficient	
κ	coupling coefficient [IEC]	
κ	electrolytic conductivity [British]	
κ	electrolytic conductivity [IUPAC]	
κ	extinction coefficient [IUPAC]	
κ	extinction coefficient [IUPAP]	
κ	magnetic susceptibility, volume [British]	
κ	Poisson's constant	
κ	ratio of specific heats [IUPAC]	
κ	ratio of specific heats [IUPAP]	
κ	specific conductance [British]	
κ	specific conductance [IUPAC]	
κ	susceptibility	
κ	susceptibility [IEC]	
κ	thermal diffusivity [British]	
κ	turbulence wave number [British]	
κ	volume viscosity	
λ	absolute activity [British]	
λ	absolute activity [IUPAC]	
λ	decay constant	
λ	decay constant [British]	
λ	decay constant [IUPAP]	
λ	disintegration constant	
λ	free path	
λ	linear charge density	
λ	linear density	
λ	longitude	
λ	mass per unit length	
λ	mean free path [British]	
λ	mobility	
λ	monochromatic (subscript)	
λ	radiation wavelength	
λ	Rossby radius of deformation	
λ	spectral (subscript)	

DPMA

λ	thermal conductivity	
λ	thermal conductivity [British]	
λ	thermal conductivity [IUPAC]	
λ	thermal conductivity [IUPAP]	
λ	wavelength	
λ	wavelength [British]	
λ	wavelength [ISO]	
λ	wavelength [IUPAC]	
λ	wavelength [IUPAP]	
λ	wavelength (subscript)	
$\bar{\lambda}$	mean free path	
λ_b	biological decay constant	
λ_B	absolute activity of substance B [IUPAP]	
λ_C	Compton wavelength [IUPAP]	
λ_e	effective wavelength	
λ_r	radiological decay constant	
λ_T	Tait free path	
λ, ϕ, r	spherical coordinates	
Λ	equivalent conductance of electrolyte [IUPAC]	
Λ	equivalent conductance of ion [IUPAC]	
Λ	equivalent conductivity	
Λ	equivalent ionic conductance [British]	
Λ	logarithmic decrement [ISO]	
Λ	loudness level [IUPAP]	
Λ	molar conductance of electrolyte [IUPAC]	
Λ	molar conductance of ion [IUPAC]	
Λ	nuclear dissociation energy	
Λ	permeance	
Λ	permeance [British]	
Λ	permeance [IEC]	
Λ^+	equivalent conductance of a positive ion [IUPAC]	
Λ^-	equivalent conductance of a negative ion [IUPAC]	
Λ°	lambda particle	
μ	absolute permeability	
μ	absolute viscosity	
μ	absorption coefficient [British]	
μ	amplification factor	
μ	chemical potential	
μ	chemical potential [British]	
μ	chemical potential [IUPAC]	
μ	chemical potential [IUPAP]	
μ	coefficient of diffusion	

μ	coefficient of friction	
μ	coefficient of friction [British]	
μ	coefficient of kinetic friction	
μ	coefficient of molecular viscosity	
μ	coefficient of sliding friction	
μ	coefficient of sliding friction [British]	
μ	dipole moment [IUPAC]	
μ	dipole moment, electric [British]	
μ	**dipole moment, magnetic [British]**	
μ	dynamic viscosity	
μ	dynamic viscosity [British]	
μ	electric moment of atom	
μ	electric moment of dipole	
μ	electric moment of molecule	
μ	electromagnetic moment [IUPAP]	
μ	gas amplification factor	
μ	Gibbs function, partial molal	
μ	grid control ratio	
μ	inductivity	
μ	ionic strength [IUPAC]	
μ	Joule-Thomson coefficient	
μ	Joule-Thomson coefficient [British]	
μ	Joule-Thomson coefficient [IUPAC]	
μ	Joule-Thomson coefficient [IUPAP]	
μ	linear absorption coefficient	
μ	linear absorption coefficient [British]	
μ	linear absorption coefficient [IUPAP]	
μ	Mach angle	
μ	Mach angle [British]	
μ	magnetic moment	
μ	magnetic moment of atom	
μ	magnetic moment of dipole	
μ	magnetic moment of molecule	
μ	magnetic moment of particle [IUPAP]	
μ	magnetic permeability	
μ	magnetic permeability [IUPAC]	
μ	micron	
μ	**molecular conductivity**	
μ	muon	
μ	muon [IUPAP]	
μ	**permeability**	
μ	permeability [British]	

μ	permeability [IEC]	
μ	permeability [IUPAP]	
μ	Poisson's ratio	
μ	reduced mass	
μ	reduced mass [IUPAP]	
μ	refractive index	
μ	refractive index [British]	
μ	relative magnetic permeability	
μ	shear elasticity	
μ_0	Bohr magneton	
μ_0	orbital magneton	
μ_0	permeability of free space [British]	
μ_0	permeability of free space [IEC]	
μ_0	permeability of vacuum [IUPAP]	
μ_0	reciprocal permeability of free space	
μ_a	atomic absorption coefficient [IUPAP]	
μ_B	Bohr magneton [IUPAP]	
μ_e	electric moment of atom	
μ_e	electric moment of dipole	
μ_e	electric moment of molecule	
μ_e	magnetic moment of electron [IUPAP]	
μ_g	refractive index, group	
μ_I	nuclear magneton	
μ_k	coefficient of kinetic friction	
μ_k	coefficient of sliding friction	
μ_l	linear absorption coefficient [IUPAP]	
μ_m	magnetic moment of atom	
μ_m	magnetic moment of dipole	
μ_m	magnetic moment of molecule	
μ_m	mass absorption coefficient [IUPAP]	
μ_n	magnetic moment of neutron [IUPAP]	
μ_N	nuclear magneton [IUPAP]	
μ_p	magnetic moment of proton [IUPAP]	
μ_r	coefficient of rolling friction	
μ_r	relative magnetic permeability	
μ_r	permeability relative [British]	
μ_r	relative permeability [IUPAP]	
μ_s	coefficient of starting friction	
μ_s	coefficient of static friction	
μ	electromagnetic moment [IUPAP]	
ν	amount of substance [IUPAP]	
ν	coefficient of kinematic viscosity	

ν	frequency	
ν	frequency [British]	
ν	frequency [IEC]	
ν	frequency [ISO]	
ν	frequency [IUPAC]	
ν	frequency [IUPAP]	
ν	kinematic viscosity	
ν	kinematic viscosity [British]	
ν	kinematic viscosity [IUPAC]	
ν	kinematic viscosity [IUPAP]	
ν	neutrino	
ν	neutrino [IUPAP]	
ν	Poisson's ratio	
ν	Poisson's ratio [British]	
ν	radiation frequency	
ν	reciprocal of dispersive power	
ν	reciprocal permeability	
ν	reluctivity	
ν	reluctivity [British]	
ν	stoichiometric number of molecules [British]	
ν	stoichiometric number of molecules [IUPAC]	
ν	stoichiometric number of molecules in a chemical reaction	
ν	wave number [British]	
ν	wave number [IUPAC]	
$\bar{\nu}$	anti-neutrino [IUPAP]	
$\bar{\nu}$	wave number [British]	
$\bar{\nu}$	wave number [ISO]	
$\bar{\nu}$	wave number [IUPAP]	
ν_0	photoelectric threshold frequency	
ν_{max}	maximum frequency	
ξ	component of particle displacement in the x direction	
ξ	displacement of sound-bearing particle	
ξ	east-west component of horizontal relative vorticity	
ξ	extent of chemical reaction	
ξ	extent of chemical reaction [IUPAC]	
ξ	extent of reaction [IUPAP]	
ξ	longitudinal displacement	
ξ	particle displacement	
Ξ	propagation flux density	

π	pion [IUPAP]	
π	ratio of circumference to diameter of circle [British]	
π	ratio of circumference to diameter of circle [IEC]	
π	ratio of the circumference of a circle to its diameter	
Π	Hertzian vector	
Π	osmotic pressure	
Π	osmotic pressure [British]	
Π	osmotic pressure [IUPAC]	
Π	osmotic pressure [IUPAP]	
Π	Peltier coefficient	
Π	Poynting vector	
Π	product [IUPAP]	
ρ	absolute humidity	
ρ	atmospheric density	
ρ	charge density [IUPAC]	
ρ	charge density [IUPAP]	
ρ	charge density, volume [British]	
ρ	density	
ρ	density [British]	
ρ	density [IEC]	
ρ	density [IUPAC]	
ρ	density [IUPAP]	
ρ	density of the medium, instantaneous	
ρ	electrical resistivity	
ρ	fraction of incident radiant power which is reflected [IUPAC]	
ρ	linear density	
ρ	mass density	
ρ	mass density [British]	
ρ	mass per unit area	
ρ	mass per unit length	
ρ	mass per unit volume	
ρ	radiant reflectance	
ρ	radius of curvature	
ρ	reflectance	
ρ	reflectance [British]	
ρ	reflection factor	
ρ	reflection factor [British]	
ρ	reflection factor [IUPAC]	
ρ	reflection factor [IUPAP]	
ρ	resistivity	

ρ	resistivity [British]	
ρ	resistivity [IEC]	
ρ	resistivity [IUPAC]	
ρ	resistivity [IUPAP]	
ρ	specific resistance	
ρ	specific resistance [British]	
ρ	surface density	
ρ	vapor density	
ρ	volume density of charge [IEC]	
ρ	volume density of electric charge	
ρ'	reflectivity	
ρ_0	density of the medium, static	
ρ_c	liquid water content per unit volume	
$\rho_\bullet$	density of air at sea level	
σ	area	
σ	cavitation number [British]	
σ	charge density, surface [British]	
σ	collision diameter of molecule	
σ	conductivity	
σ	conductivity [British]	
σ	conductivity [IEC]	
σ	conductivity [IUPAP]	
σ	cross-contraction ratio	
σ	**cross-section**	
σ	**cross-section [British]**	
σ	**cross-section [IUPAP]**	
σ	density ratio	
σ	**diameter of molecule [IUPAC]**	
σ	**dispersion**	
σ	effective molecular cross-section	
σ	electric conductivity	
σ	entropy production rate	
σ	leakage coefficient [IEC]	
σ	magnetic leakage coefficient	
σ	mass per unit area	
σ	normal stress	
σ	normal stress [British]	
σ	normal stress [IUPAP]	
σ	Poisson's ratio	
σ	Poisson's ratio [British]	
σ	Prandtl number	
σ	slip (electrical machinery)	

σ	specific conductance [British]	
σ	specific magnetization	
σ	standard deviation	
σ	standard deviation of a distributed variate [British]	
σ	Stefan-Boltzmann constant	
σ	Stefan-Boltzmann constant [British]	
σ	surface charge density	
σ	surface charge density [IUPAP]	
σ	surface density	
σ	surface density of charge [IEC]	
σ	surface tension	
σ	surface tension [British]	
σ	surface tension [IUPAC]	
σ	surface tension [IUPAP]	
σ	symmetry number [IUPAC]	
σ	Thomson coefficient	
σ	traction [IUPAC]	
σ	wave number	
σ	wave number [ISO]	
σ	wave number [IUPAC]	
σ	wave number [IUPAP]	
σ^2	variance	
σ_a	absorption cross-section	
σ_s	scattering cross-section	
Σ	exposure	
Σ	macroscopic cross-section [British]	
Σ	macroscopic cross-section [IUPAP]	
Σ	summation	
Σ	summation [IEC]	
Σ	summation [IUPAP]	
Σ_a	absorption cross-section	
Σ_s	scattering cross-section	
τ	decay modulus	
τ	dew-point temperature	
τ	fraction of incident radiant power which is transmitted [IUPAC]	
τ	growth, period of [British]	
τ	mean-life	
τ	mean-life [IUPAP]	
τ	optical transmittance	
τ	period	

τ		periodic time [IUPAC]
τ		period of decay [British]
τ		radiant transmittance
τ		relaxation time
τ		shear stress
τ		shear stress [British]
τ		shear stress [IUPAC]
τ		shear stress [IUPAP]
τ		shear stress in fluid [British]
τ		time
τ		time constant
τ		time constant [IEC]
τ		time constant of an exponentially varying quantity [ISO]
τ		time interval [IUPAC]
τ		total volume
τ		transmission factor
τ		transmission factor [British]
τ		transmission factor [IUPAC]
τ		transmission factor [IUPAP]
τ		transmissivity
τ		transmittance
τ		transmittance [British]
$\vec{\tau}$		unit vector tangent to path
Υ		admittance [IUPAC]
φ		angle of incidence
φ		argument of z [IUPAP]
φ		electromagnetic scalar potential
φ		electron affinity
φ		epoch angle
φ		fluidity
φ		fluidity [IUPAC]
φ		net work function per unit charge
φ		osmotic coefficient [IUPAC]
φ		osmotic coefficient [IUPAP]
φ		phase angle
φ		phase difference [IEC]
φ		plane angle [IUPAP]
φ		reciprocal of viscosity
φ		scattering angle [IUPAP]
φ		velocity potential
$\bar{\varphi}$		polarizing angle of a dielectric material

$\bar{\varphi}$	principal angle of incidence	
φ'	angle of refraction	
φ_B	volume fraction of substance B [IUPAP]	
φ_c	critical angle	
φ_g	gross electron affinity	
φ_g	gross work function per unit charge	
φ_{ij}	interaction energy between molecules i and j [IUPAP]	
ϕ	angle of friction [British]	
ϕ	electronic exit work function [British]	
ϕ	fluidity [British]	
ϕ	geopotential	
ϕ	heat flux intensity [British]	
ϕ	phase angle	
ϕ	phase difference [British]	
ϕ	phase difference [IEC]	
ϕ	relative humidity [British]	
ϕ	shear strain [British]	
ϕ	velocity potential	
ϕ	velocity potential [British]	
Φ	electric potential [IUPAP]	
Φ	light flux [IUPAC]	
Φ	luminous flux [British]	
Φ	luminous flux [IUPAP]	
Φ	magnetic flux	
Φ	magnetic flux [British]	
Φ	magnetic flux [IEC]	
Φ	magnetic flux [IUPAP]	
Φ	radiant flux	
Φ	radiant power	
Φ	radiant power [IUPAC]	
Φ_e	radiant flux [IUPAP]	
χ	magnetic susceptibility [IUPAC]	
χ	magnetic susceptibility, mass [British]	
χ	magnetic susceptibility, specific	
χ	susceptibility	
χ_e	electric susceptibility	
χ_e	electric susceptibility [IUPAP]	
χ_m	magnetic susceptibility	
χ_m	magnetic susceptibility [IUPAP]	
ψ	azimuth angle	
ψ	phase angle	

ψ		stream function
ψ		stream function [British]
ψ		time-dependent wave function
$\bar{\psi}$		principal azimuth angle
Ψ		electric flux [British]
Ψ		electric flux [IEC]
Ψ		electric flux of induction
Ψ		Planck's function
Ψ		total dielectric flux
Ψ		total electric flux
ω		angular frequency
ω		angular frequency [British]
ω		angular frequency [IEC]
ω		angular frequency [ISO]
ω		angular frequency [IUPAC]
ω		angular frequency [IUPAP]
ω		angular frequency of impressed force
ω		angular frequency without damping
ω		angular speed
ω		angular velocity
ω		angular velocity [British]
ω		angular velocity [IEC]
ω		angular velocity [ISO]
ω		angular velocity [IUPAC]
ω		angular velocity [IUPAP]
ω		circular frequency
ω		dispersive power
ω		periodicity
ω		precession rate
ω		pulsatance
ω		pulsatance [IUPAP]
ω		solid angle
ω		solid angle [British]
ω		solid angle [IEC]
ω		solid angle [ISO]
ω		solid angle [IUPAC]
ω		solid angle [IUPAP]
ω		specific magnetic rotation
ω		Verdet constant
ω		vorticity [British]
ω'		angular frequency of free vibration with damping
ω_L		Larmor (angular) frequency [IUPAP]

ω_r	....	resonant periodicity ....
ω	....	angular velocity ....
Ω	....	angular speed of rotation of Earth ....
Ω	....	angular velocity ....
Ω	....	angular velocity [British] ....
Ω	....	load factor [British] ....
Ω	....	solid angle ....
Ω	....	solid angle [British] ....
Ω	....	solid angle [IEC] ....
Ω	....	solid angle [ISO] ....
Ω	....	solid angle [IUPAP] ....
Ω	....	volume in gamma phase space [IUPAP] ....
Ω	....	volume of phase space ....

LETTER SYMBOLS

Alphabetically by Definition

absolute activity [British] .... λ
absolute activity [IUPAC] .... λ
absolute activity of substance B [IUPAP] .... λ_B
absolute humidity .... ρ
absolute permeability .... μ
absolute temperature .... T
absolute temperature .... Θ
absolute temperature [British] .... T
absolute temperature [IEC] .... T
absolute temperature [IEC] .... Θ
absolute temperature [IUPAC] .... T
absolute temperature of ice point .... T_0
absolute virtual temperature .... T^*
absolute virtual temperature .... T_v
absolute viscosity .... μ
absolute vorticity, vertical component of ...... ζ_a
absorbance (light) [IUPAC] .... A
absorbance (light) [IUPAC] .... E
absorbance, specific [IUPAC] .... a
absorptance .... α
absorption coefficient [British] .... μ
absorption coefficient [IUPAC] .... κ
absorption coefficient [IUPAP] .... a
absorption coefficient, atomic [IUPAP] .... μ_a
absorption coefficient, linear [IUPAP] .... μ
absorption coefficient, linear [IUPAP] .... μ_l
absorption coefficient, mass [IUPAP] .... μ_m
absorption, coefficient of .... α
absorption coefficient of absorbing material ... α
absorption cross-section .... σ_a
absorption cross-section .... Σ_a
absorption factor .... α
absorption factor [British] .... α
absorption factor [IUPAC] .... α
absorption factor [IUPAP] .... α
absorption factor, acoustic [IUPAP] .... α
absorption factor, acoustic [IUPAP] .... α_a
absorption index .... κ

absorption, molar decadic [IUPAC]	ϵ
absorption, radiant	α
absorptivity	α
absorptivity [IUPAC]	a
absorptivity, molar [IUPAC]	ϵ
acceleration	a
acceleration [ISO]	a
acceleration [IUPAC]	a
acceleration [IUPAP]	a
acceleration, angular	α
acceleration, angular	$\boldsymbol{\alpha}$
acceleration, angular [British]	α
acceleration, angular [ISO]	α
acceleration, angular [IUPAP]	α
acceleration due to gravity	g
acceleration due to gravity [British]	g
acceleration due to gravity [IEC]	g
acceleration due to gravity [ISO]	g
acceleration due to gravity [IUPAP]	g
acceleration due to gravity, local	g_L
acceleration due to gravity, standard [British]	g_n
acceleration, linear	$\mathbf{a}$
acceleration, linear [British]	a
acceleration, linear [British]	f
acceleration, linear [IEC]	a
acceleration of free fall	g
acceleration of free fall [IUPAC]	g
acceleration, standard gravitational [IUPAP]	g_n
accommodation, coefficient of	a
acoustic absorption factor [IUPAP]	α
acoustic absorption factor [IUPAP]	α_a
acoustical admittance	y_A
acoustical admittance	Y_A
acoustical conductance	g_A
acoustical conductance	G_A
acoustical impedance, complex	z_A
acoustical impedance, complex	Z_A
acoustical power	P_A
acoustical reactance	x_A
acoustical reactance	X_A
acoustical resistance	r_A
acoustical resistance	R_A

DPMA

acoustical resistance, specific	r_A
acoustical response	R_A
acoustical susceptance	b_A
acoustical susceptance	B_A
acoustic conductivity of an opening	c
acoustic intensity	I
acoustic source, strength of simple	A
action	S
action variable	J
activation energy [British]	E
active power	P
active power [IEC]	P
activity [IUPAP]	A
activity, absolute [British]	λ
activity, absolute [IUPAC]	λ
activity at specified time (radioactivity)	I
activity, chemical	a
activity coefficient [British]	f
activity coefficient [British]	γ
activity coefficient [IUPAC]	f
activity coefficient [IUPAC]	γ
activity coefficient of substance B [IUPAP]	γ_B
activity coefficient, stoichiometric	γ
activity coefficient, stoichiometric [British]	γ
activity, initial (radioactivity)	I_0
activity of substance B [IUPAP]	z_B
activity of substance X, relative [British]	a_x
activity of substance X, relative [British]	$\{X\}$
activity, relative [British]	a
activity, relative [IUPAC]	a
actual lapse rate of temperature	γ
adiabatic (subscript)	ad
adiabatic (subscript)	s
adiabatic exponent	γ
adiabatic wall (subscript)	aw
admittance	Y
admittance [British]	Y
admittance [IEC]	Y
admittance [IUPAC]	Y
admittance [IUPAP]	Y
admittance, acoustical	y_A
admittance, acoustical	Y_A

admittance, electrical	y_E
admittance, electrical	Y_E
admittance, mechanical	y_M
admittance, mechanical	Y_M
admittance, output	y_p
admittance, plate	y_p
admittance, rotational	y_R
admittance, rotational	Y_R
admittance with plate load, grid	y_g
adsorbed (subscript)	a
affinity	A
affinity [IUPAP]	A
affinity, electron	φ
affinity, gross electron	φ_g
affinity of a chemical reaction [IUPAC]	A
aggregate quantity of heat [British]	Q
aging, coefficient of (piezoelectricity)	c
air at sea level, density of	ρ_0
albedo	A
alpha particle	α
alpha particle [IUPAP]	α
altitude	h
altitude	y
altitude	z
altitude (subscript)	h
altitude, pressure	z_p
ambient (subscript)	a
amount of illumination	E
amount of substance [IUPAP]	n
amount of substance [IUPAP]	ν
amplification factor	μ
amplification factor, gas	μ
amplification of amplifier, power	A
amplification of amplifier, power	A_p
amplification of amplifier, voltage	A
amplification of amplifier, voltage	A_v
amplitude	A
amplitude of simple harmonic pressure	P
amplitude of velocity potential	A
angle	α
angle, azimuth	ψ
angle between ray and normal in first medium	i

DPMA

angle between ray and normal in first medium	φ
angle between ray and normal in second medium	r
angle between ray and normal in second medium	φ'
angle, contact	θ
angle, contact [IUPAC]	θ
angle, critical	φ_c
angle, epoch	ϵ
angle, epoch	φ
angle, glancing	θ
angle, Mach	μ
angle, Mach [British]	μ
angle, nadir	η
angle, normal	θ
angle of a dielectric material, polarizing	$\bar{\varphi}$
angle of deviation	δ
angle of diffraction	θ
angle of friction [British]	ϕ
angle of incidence	i
angle of incidence	φ
angle of incidence, principal	$\bar{\varphi}$
angle of minimum deviation	D
angle of minimum deviation	δ_m
angle of optical rotation	α
angle of optical rotation [IUPAC]	α
angle of prism, refracting	A
angle of reflection	r
angle of refraction	r
angle of refraction	φ'
angle of slope in image space	u'
angle of slope in object space	u
angle of twist [British]	θ
angle, phase	θ
angle, phase	φ
angle, phase	ϕ
angle, phase	ψ
angle, plane	θ
angle, principal azimuth	$\bar{\psi}$
angle, solid	ω
angle, solid [British]	ω
angle, solid [British]	Ω
angle, solid [IEC]	Ω
angle, solid [ISO]	ω

DPMA

angle, solid [ISO]	Ω
angle, solid [IUPAC]	ω
angle, solid [IUPAP]	ω
angle, solid [IUPAP]	Ω
angular acceleration	α
angular acceleration	α
angular acceleration [British]	α
angular acceleration [ISO]	α
angular acceleration [IUPAP]	α
angular dispersion	D
angular displacement	θ
angular distance	θ
angular frequency	ω
angular frequency [British]	ω
angular frequency [IEC]	ω
angular frequency [ISO]	ω
angular frequency [IUPAC]	ω
angular frequency [IUPAP]	ω
angular frequency of free vibration with damping	n'
angular frequency of free vibration with damping	ω'
angular frequency of impressed force	p
angular frequency of impressed force	ω
angular frequency without damping	n
angular frequency without damping	ω
angular magnification	γ
angular momentum, intrinsic	s
angular momentum, specific	h
angular momentum, total	H
angular resolving power of telescope	α
angular speed	ω
angular speed of rotation of Earth	Ω
angular velocity	ω
angular velocity	ω
angular velocity	Ω
angular velocity [British]	ω
angular velocity [British]	Ω
angular velocity [IEC]	ω
angular velocity [ISO]	ω
angular velocity [IUPAC]	ω
angular velocity [IUPAP]	ω
anode terminal	A
anti-neutrino [IUPAP]	$\bar{\nu}$

DPMA

antiresonant frequency	f_A
aperture	a
apparent power	P_s
apparent power [British]	S
apparent power [IEC]	P_s
apparent power [IEC]	S
arc length	s
arc length [IUPAC]	s
arc of curve [IEC]	s
area	A
area	S
area	v
area [British]	A
area [IEC]	A
area [IEC]	S
area [ISO]	A
area [ISO]	S
area [IUPAC]	A
area [IUPAC]	S
area [IUPAP]	A
area [IUPAP]	S
area, cross-sectional	A
area, cross-sectional [British]	A
area for drag and lift, reference	A
area, major hydraulic	A
area of a diaphragm	S
area of a radiator	S
area of a tube	S
area of cross-section of walls	S
area of equatorial plane of Earth	F
area, surface	A
area, surface, per unit volume	a
argument of z [IUPAP]	φ
arithmetic (subscript)	a
aspect ratio	A
aspect ratio [British]	A
atmospheric pressure	p
atmospheric pressure	p_a
atmospheric pressure [British]	p_a
atmospheric pressure [British]	p_{at}
atmospheric pressure [British]	P_{at}
atomic absorption coefficient [IUPAP]	μ_a

atomic mass [IUPAP]	M
atomic mass [IUPAP]	M_a
atomic mass constant, unified [IUPAP]	m_u
atomic mass, relative [IUPAP]	A_r
atomic number	Z
atomic number [British]	Z
atomic number [IUPAP]	Z
atomic stopping power [IUPAP]	S_a
atomic volume	V
atomic weight	A
atomic weight	M
atomic weight [British]	A
attenuation coefficient [British]	α
attenuation coefficient [IEC]	a
attenuation coefficient [IEC]	α
attenuation coefficient [ISO]	α
attenuation constant	α
attenuation, optical	D
attraction energy, molecular [IUPAP]	ϵ
Austausch coefficient	A
average binding energy per nucleon in a nucleus [British]	$\bar{B}$
average current	$\bar{I}$
average energy expended in a gas per ion pair formed	W
average molecular kinetic energy	ϵ
average particle velocity	u_a
average sound pressure	p_a
average speed	$\bar{u}$
average speed	u_{av}
average speed	$\bar{v}$
average speed	v_{av}
average value of a quantity x	$\bar{x}$
average velocity [IUPAP]	$\bar{c}$
average velocity [IUPAP]	c_0
average velocity [IUPAP]	$\bar{u}$
average velocity [IUPAP]	u_0
Avogadro's constant [IUPAP]	L
Avogadro's constant [IUPAP]	N_A
Avogadro's number	N
Avogadro's number	N_0
Avogadro's number [British]	N

Avogadro's number [IUPAC]	L
Avogadro's number [IUPAC]	N
Avogadro's number [IUPAC]	N_0
axial vector coupling [IUPAP]	A
azimuth angle	ψ
azimuth angle, principal	$\bar{\psi}$
barometric (subscript)	bar
base (subscript)	b
base of Naperian logarithms	e
base of Naperian logarithms	ϵ
base of natural logarithms [British]	e
base of natural logarithms [IEC]	e
base of natural logarithms [IEC]	ε
base of natural logarithms [IEC]	ϵ
base of natural logarithms [IUPAP]	e
bending moment	M
bending moment [British]	M
Bessel equation, root of a	η_1
bias or remanent displacement	D_0
biological decay constant	λ_b
black body (subscript)	b
black body (subscript)	B
Bohr magneton	μ_0
Bohr magneton [British]	β
Bohr magneton [IUPAP]	β
Bohr magneton [IUPAP]	μ_B
Bohr magneton, nuclear	μ_I
Bohr radius	a_1
Bohr radius [IUPAP]	a_0
Boltzmann's constant	k
Boltzmann's constant [British]	k
Boltzmann's constant [IUPAC]	k
Boltzmann's constant [IUPAC]	$\Bbbk$
Boltzmann's constant [IUPAP]	k
Boltzmann's function	H
Boltzmann's function [IUPAP]	H
boundary-layer displacement thickness, in two-dimensional and axi-symmetric flow [British]	δ^*
boundary-layer displacement thickness, in two-dimensional and axi-symmetric flow [British]	δ_1

boundary-layer energy thickness, in two-dimensional and axi-symmetric flow [British]	δ_3
boundary-layer momentum thickness, in two-dimensional and axi-symmetric flow [British]	δ_2
boundary-layer momentum thickness, in two-dimensional and axi-symmetric flow [British]	θ
boundary layer, momentum thickness of	θ
boundary-layer shape parameter, in two-dimensional and axi-symmetric flow [British]	H
boundary-layer thickness	δ
boundary-layer thickness [British]	δ
Bragg planes in a crystal, spacing of	d
breadth	b
breadth [British]	b
breadth [IEC]	b
breadth [ISO]	b
breadth [IUPAP]	b
brightness	B
brightness, photometric	B
brightness, photometric [British]	L
bulk modulus [British]	K
bulk modulus [IUPAP]	K
bulk modulus of elasticity	K
calculated (subscript)	calc
calibrated (subscript)	cal
candlepower	I
capacitance	C
capacitance [British]	C
capacitance [IEC]	C
capacitance [IUPAP]	C
capacitance, acoustical	C_A
capacitance coefficient, partial	Q
capacitance constant, shunt	Γ_0
capacitance, electrical	C_E
capacitance, partial	c
capacitance ratio	r
capacitance, reciprocal	S
capacitive reactance	X_C
capacitivity, in centimeter-gram-second units	ϵ
capacitivity, in meter-kilogram-second units	ϵ_r
capacity [IUPAC]	C

capacity, specific inductive	ϵ
capacity, specific inductive	ϵ_r
cathode resistance	R_k
cathode terminal	K
Cauchy constant	A
Cauchy constant	B
Cauchy constant	C
cavitation number [British]	σ
characteristic impedance	Z_0
characteristic temperature [IUPAC]	Θ
characteristic temperature [IUPAP]	Θ
characteristic temperature, Debye	θ_D
characteristic temperature, Debye	Θ
charge density [IUPAC]	ρ
charge density [IUPAP]	ρ
charge density, linear	λ
charge density, surface	σ
charge density, surface [British]	σ
charge density, surface [IUPAP]	σ
charge density, volume [British]	ρ
charge, electric	q
charge, electric	Q
charge, electric [British]	Q
charge, electronic	$-e$
charge, elementary [IUPAC]	e
charge, elementary [IUPAC]	e
charge, instantaneous	q
charge, nuclear	Z
charge number of an ion [IUPAC]	z
charge number of ion [IUPAP]	z
charge of electron [British]	$-e$
charge of positron [IUPAP]	e
charge, surface density of [IEC]	σ
charge, volume density of [IEC]	ρ
chemical potential	μ
chemical potential [British]	μ
chemical potential [IUPAC]	μ
chemical potential [IUPAP]	μ
chemical reaction, extent of	ξ
Chezy coefficient	C
Chezy coefficient [British]	C
chord	c

Term	Symbol
chord [British]	c
chord (subscript)	c
circle of least confusion, radius of	Z
circular frequency	ω
circular wave number	n
circular wave number [ISO]	k
circular wave number [IUPAP]	k
circulation	C
circulation	Γ
circulation [British]	K
circulation [British]	Γ
cloud (subscript)	c
cloud amount	N
coefficient	c
coefficient	C
coefficient	k
coefficient [British]	C
coefficient, absorption [IUPAC]	κ
coefficient, activity [IUPAC]	f
coefficient, activity [IUPAC]	γ
coefficient, atomic absorption [IUPAP]	μ_a
coefficient, attenuation [IEC]	α
coefficient, attenuation [ISO]	α
coefficient, Austausch	A
coefficient, Chezy	C
coefficient, coupling [IEC]	κ
coefficient, cubic expansion [IUPAP]	γ
coefficient, damping	δ
coefficient, extinction	ϵ
coefficient, extinction [IUPAC]	κ
coefficient, extinction [IUPAP]	κ
coefficient, induction	c
coefficient, Joule-Thomson	μ
coefficient, Joule-Thomson [British]	μ
coefficient, Joule-Thomson [IUPAC]	μ
coefficient, Joule-Thomson [IUPAP]	μ
coefficient, leakage [IEC]	σ
coefficient, linear absorption	μ
coefficient, linear absorption [British]	μ
coefficient, linear absorption [IUPAP]	μ
coefficient, linear absorption [IUPAP]	μ_l
coefficient, magnetic leakage	σ

DPMA

coefficient, mass absorption [IUPAP] .... μ_m
coefficient, molar extinction [IUPAC] .... ϵ
coefficient of absorption .... α
coefficient of accommodation .... a
coefficient of aging (piezoelectricity) c
coefficient of compressibility .... k
coefficient of diffusion .... μ
coefficient of diffusion [IUPAC] .... D
coefficient of drag .... C_D
coefficient of drag [British] .... C_D
coefficient of eddy transfer, dynamic .... A
coefficient of eddy transfer, kinematic .... K
coefficient of friction .... f
coefficient of friction .... μ
coefficient of friction [British] .... μ
coefficient of friction [IUPAC] .... f
coefficient of friction [IUPAP] .... f
coefficient of heat transfer [British] .... h
coefficient of heat transfer [British] .... α
coefficient of kinematic viscosity .... ν
coefficient of kinetic friction .... μ
coefficient of kinetic friction .... μ_k
coefficient of linear expansion .... α
coefficient of molecular viscosity .... μ
coefficient of pressure [British] .... C_p
coefficient of recombination .... α
coefficient of resilience .... e
coefficient of resistance, temperature .... α
coefficient of restitution .... e
coefficient of rolling friction .... μ_r
coefficient of sliding friction .... f
coefficient of sliding friction .... μ
coefficient of sliding friction .... μ_k
coefficient of sliding friction [British] .... μ
coefficient of starting friction .... μ_s
coefficient of static friction .... μ_s
coefficient of thermal expansion, cubic [British] ... γ
coefficient of thermal expansion, linear [British] ... α
coefficient of viscosity .... η
coefficient, osmotic [IUPAC] .... g
coefficient, osmotic [IUPAC] .... φ
coefficient, over-all heat transfer .... U

coefficient, over-all heat transfer [British]	U
coefficient, partial capacitance	c
coefficient, Peltier	Π
coefficient, propagation [British]	γ
coefficient, propagation [IEC]	p
coefficient, propagation [IEC]	γ
coefficient, propagation [ISO]	γ
coefficient, relative dielectric	ϵ
coefficient, relative dielectric	ϵ_r
coefficient, scattering	s
coefficient, stoichiometric activity	γ
coefficient, stoichiometric activity [British]	γ
coefficient, Thomson	σ
coefficient, turbulence exchange	ϵ
coercive force	H_c
colatitude	θ
collision diameter of molecule	σ
collision frequency, molecular	Z
complex accoustical impedance	Z_A
complex accoustical impedance	z_A
complex conjugate of A [IUPAP]	A^*
complex conjugate of z [IUPAP]	$\bar{z}$
complex conjugate of z [IUPAP]	z^*
complex electrical impedance	Z_E
complex electrical impedance	z_E
complex mechanical impedance	z_M
complex mechanical impedance	Z_M
complex rotational impedance	z_R
complex rotational impedance	Z_R
complex velocity potential [British]	w
compliance	C
compliance, mechanical	C_M
compliance, rotational	C_R
component A (subscript)	A
component B (subscript)	B
component C (subscript)	C
component of particle displacement in the x direction	ξ
component of particle displacement in the y direction	η
component of particle displacement in the z direction	ζ
component of particle velocity in x direction	u

DPMA

component of particle velocity in y direction	v
component of particle velocity in z direction	w
compressibility [British]	κ
compressibility [IUPAC]	κ
compressibility [IUPAP]	κ
compressibility, coefficient of	k
compressibility factor	k
compressibility factor, subsonic [British]	β
compressibility factor, supersonic [British]	B
compressibility factor, supersonic [British]	β
compression modulus [IUPAC]	K
Compton wavelength [IUPAP]	λ_C
concentrated force	F
concentrated force	P
concentrated force	Q
concentrated load	F
concentrated load	P
concentrated load	Q
concentration	c
concentration [British]	c
concentration [British]	C
concentration [IUPAC]	c
concentration at zero degree centigrade and one atmosphere, molecular	n_0
concentration, mass	c
concentration, molecular	n
concentration, molecular [IUPAC]	C
concentration of substance B, molar	c_B
concentration of substance X, molar [British]	c_x
concentration of substance X, molar [British]	C_x
concentration of substance X, molar [British]	$[X]$
concentration, surface [British]	Γ
concentration, surface [IUPAC]	Γ
condensation	s
condensation level (subscript)	c
condensation processes (subscript)	c
conductance	g
conductance	G
conductance [British]	G
conductance [IEC]	G
conductance [IUPAP]	G
conductance, acoustical	g_A
conductance, acoustical	G_A

conductance at zero frequency and constant grid potential, plate ... k_p
conductance at zero frequency and constant plate potential, grid ... k_g
conductance, electrical ... g_E
conductance, electrical ... G_E
conductance, input ... g_g
conductance, mechanical ... g_M
conductance, mechanical ... G_M
conductance, mutual ... G_M
conductance of a negative ion, equivalent [IUPAC] ... Λ^-
conductance of a positive ion, equivalent [IUPAC] ... Λ^+
conductance of electrolyte, equivalent [IUPAC] ... Λ
conductance of electrolyte, molar [IUPAC] ... Λ
conductance of ion, equivalent [IUPAC] ... Λ
conductance of ion, molar [IUPAC] ... Λ
conductance, output ... g_p
conductance, plate ... g_p
conductance, rotational ... g_R
conductance, rotational ... G_R
conductance, specific [British] ... γ
conductance, specific [British] ... κ
conductance, specific [British] ... σ
conductance, specific [IUPAC] ... κ
conductance, thermal ... C
conduction current ... I
conductivity ... γ
conductivity ... σ
conductivity [British] ... γ
conductivity [British] ... σ
conductivity [IEC] ... γ
conductivity [IEC] ... σ
conductivity [IUPAP] ... γ
conductivity [IUPAP] ... σ
conductivity, electric ... γ
conductivity, electric ... σ
conductivity, electrolytic [British] ... κ
conductivity, electrolytic [IUPAC] ... κ
conductivity, equivalent ... Λ
conductivity, molecular ... μ
conductivity of an opening, acoustic ... c
conductivity, thermal ... k

conductivity, thermal	λ
conductivity, thermal [British]	k
conductivity, thermal [British]	λ
conductivity, thermal [IUPAC]	λ
conductivity, thermal [IUPAP]	λ
configuration-space volume	V
conjugate of z [IUPAP]	z^*
constant	C
constant, Boltzmann's	k
constant, electric	Γ_e
constant electric displacement (superscript)	D
constant electric field (superscript)	E
constant energy (subscript)	E
constant enthalpy (subscript) [British]	h
constant entropy (subscript)	S
constant entropy (subscript) [British]	s
constant, magnetic	Γ_m
constant pressure (subscript)	p
constant pressure (subscript) [British]	p
constant, reaction velocity	k
constant, spring	k
constant strain (superscript)	S
constant stress (superscript)	T
constant temperature (subscript)	T
constant temperature (subscript) [British]	T
constant, torsion	k
constant volume (subscript)	v
constant volume (subscript) [British]	v
contact angle [IUPAC]	θ
contact or Volta potential	V_v
contrast of photographic emulsion	γ
control grid (subscript)	g
control ratio, grid	μ
convection (subscript) [British]	c
convection current	I
conversion loss	L_c
coordinate along X-axis	x
coordinate along Y-axis	y
coordinate along Z-axis	z
coordinates, Cartesian orthogonal	x, y, z
coordinates, Cartesian orthogonal	X, Y, Z
coordinates, spherical	λ, ϕ, r

Coriolis parameter	f
Coriolis parameter	l
Coriolis parameter, latitudinal variation rate of	β
coupling coefficient [IEC]	k
coupling coefficient [IEC]	κ
coupling factor	k
coupling factor, longitudinal (piezoelectricity)	k_{33}
coupling factor, planar (piezoelectricity)	k_p
coupling factor, shear (piezoelectricity)	k_{15}
coupling factor, thickness (piezoelectricity)	k_t
coupling factor, transverse (piezoelectricity)	k_{31}
critical angle	φ_c
critical pressure	p_c
critical properties (subscript)	c
critical state (subscript)	c
critical state (subscript) [British]	c
critical temperature	t_c
critical temperature	T_c
critical value (subscript)	c
critical value (subscript) [British]	c
critical volume	V_c
cross-contraction ratio	σ
cross-section	σ
cross-section [British]	σ
cross-section [IUPAP]	σ
cross-section, absorption	σ_a
cross-section, absorption	Σ_a
cross-sectional area	A
cross-sectional area [British]	A
cross-section, effective molecular	σ
cross-section, macroscopic [British]	Σ
cross-section, macroscopic [IUPAP]	Σ
cross-section, scattering	σ_s
cross-section, scattering	Σ_s
crystalline (subscript) [British]	X
cubic coefficient of thermal expansion [British]	γ
cubic expansion coefficient [IUPAP]	γ
Curie constant	C
Curie temperature	T_c
current	I
current [IEC]	I
current, average	I
current, average	I_{av}

DPMA

current, capacitive	I_c
current, conduction	I
current, convection	I
current density, electric	J
current density, electric [IUPAC]	J
current density, electric [IUPAP]	J
current density, electric [IUPAP]	J
current density, linear	A
current density, linear [IEC]	A
current density, linear [IEC]	α
current, displacement	$\dot{D}$
current, displacement	$\dot{D}/4\pi$
current, effective	I
current, electric [British]	I
current, electric [IUPAC]	i
current, electric [IUPAC]	I
current, electric [IUPAP]	I
current, inductive	I_L
current, instantaneous	i
current, instantaneous [British]	i
current, maximum	I_m
current, maximum peak	I_m
current, maximum peak	I_{mp}
current, peak	$\hat{I}$
current, peak	I_p
current, peak	I_{pk}
current, quiescent	I
current, root-mean-square	I
current, steady direct	I
current through resistance	I_R
current through resistance, instantaneous	i_R
curvature	K
curve, arc of [IEC]	s
curved path [IEC]	s
customary temperature [British]	t
customary temperature [British]	θ
cutoff voltage	E_{co}
damping coefficient	δ
damping coefficient [IEC]	δ
damping coefficient [ISO]	δ
damping coefficient, velocity	c

Debye characteristic temperature	θ_D
Debye characteristic temperature	Θ
Debye temperature [IUPAP]	Θ_D
decadic absorption, molar [IUPAC]	ϵ
decay constant	λ
decay constant [British]	λ
decay constant [IUPAP]	λ
decay constant, biological	λ_b
decay constant, radiological	λ_r
decay modulus	τ
decay, period of [British]	τ
decimetric solar flux	F_{10}
declination of geomagnetic field	D
decrement, logarithmic [ISO]	Λ
deflection	δ
deflection [British]	δ
deflection of beam	δ
deflection of beam, maximum	δ
deformation, Rossby radius of	λ
degeneracy	g
degree of dissociation	α
degree of dissociation [British]	α
degree of dissociation [IUPAC]	α
degree of electrolytic dissociation [IUPAC]	α
degree of hydrolysis	h
degree of ionization	α
degree of reaction	α
degree of reaction [IUPAC]	α
degrees of freedom	f
density	D
density	ρ
density [British]	ρ
density [IEC]	δ
density [IEC]	ρ
density [IUPAC]	ρ
density [IUPAP]	ρ
density, charge [IUPAC]	ρ
density, charge [IUPAP]	ρ
density, dielectric flux	**D**
density, electric flux	**D**
density, electric flux [British]	D
density, electric flux [IEC]	D
density, linear	λ

density, linear	ρ
density, linear charge	λ
density, linear current	A
density, magnetic flux	B
density, mass	ρ
density, mass [British]	ρ
density, molecular	n
density, number	n
density of air at sea level	ρ_0
density of charge, surface [IEC]	σ
density of charge, volume [IEC]	ρ
density of electric charge, volume	ρ
density of radiant energy	u
density of the medium, instantaneous	ρ
density of the medium, static	ρ_0
density, optical	D
density, optical [British]	d
density ratio	σ
density, relative [British]	d
density, relative [IUPAC]	d
density, relative [IUPAP]	d
density, surface	ρ
density, surface	σ
density, surface charge	σ
density, surface charge [British]	σ
density, surface charge [IUPAP]	σ
density, vapor	ρ
density, volume charge [British]	ρ
departure from normal (subscript)	δ
depth	h
depth	H
depth	y
depth [IEC]	h
deuteron	d
deuteron [IUPAP]	d
deviation	δ
deviation angle	δ
deviation, angle of minimum	δ_m
deviation, frequency	δ
deviation, standard	σ
dew point (subscript)	d
dew-point temperature	τ

DPMA

diameter	d
diameter	D
diameter [British]	d
diameter [British]	D
diameter [IEC]	d
diameter [ISO]	d
diameter [IUPAC]	d
diameter [IUPAP]	d
diameter of molecule [IUPAC]	D
diameter of molecule [IUPAC]	σ
diameter of molecule, collision	σ
dielectric coefficient	ϵ
dielectric coefficient, in centimeter-gram-second units	ϵ
dielectric coefficient, in meter-kilogram-second units	ϵ_r
dielectric coefficient, relative	ϵ
dielectric coefficient, relative	ϵ_r
dielectric constant [British]	ϵ
dielectric constant [IUPAC]	ϵ
dielectric flux density	$\mathbf{D}$
dielectric flux, total	Ψ
dielectric loss angle [British]	δ
dielectric polarization [IUPAC]	P
dielectric polarization [IUPAP]	P
dielectric polarization [IUPAP]	$\boldsymbol{P}$
difference in optical path	Δ
difference in phase	δ
difference, phase [British]	ϕ
difference, phase [IEC]	φ
difference, phase [IEC]	ϕ
difference, temperature	Δt
differential operator	d
differential operator [IEC]	d
diffraction, angle of	θ
diffusion coefficient [British]	D
diffusion coefficient [IUPAP]	D
diffusion, coefficient of	μ
diffusion, coefficient of [IUPAC]	D
diffusion coefficient, thermal [IUPAP]	D_T
diffusion factor, thermal [IUPAP]	α_T
diffusion ratio, thermal [IUPAP]	K_T

diffusivity, mass	D
diffusivity, thermal	α
diffusivity, thermal [British]	κ
diffusivity, volumetric	D
dilatation	Δ
dilution (subscript) [British]	d
dimensionless number	N
dioptric power	D
dip angle	δ
dipole moment [IUPAC]	μ
dipole moment, electric [British]	μ
dipole moment, electric [IUPAP]	p
dipole moment, electric [IUPAP]	$\boldsymbol{p}$
dipole moment, magnetic [British]	μ
dipole moment, magnetic [IUPAP]	j
dipole moment, magnetic [IUPAP]	$\boldsymbol{j}$
direction cosine	l
direction cosine	m
direction cosine	n
directivity index	D_i
directivity ratio	R_θ
direct solar radiant intensity, in or below atmosphere	I
direct solar radiant intensity, in or below atmosphere	J
direct strain [British]	e
direct strain [British]	ϵ
discharge rate	q
disintegration constant	λ
disintegration energy [IUPAP]	Q
dispersion	σ
dispersion, angular	D
dispersive power	ω
dispersive power, reciprocal of	ν
displacement	s
displacement	δ
displacement angle, small	δ
displacement, angular	θ
displacement component of sound-bearing particle	ζ
displacement component of sound-bearing particle	η
displacement component of sound-bearing particle	ξ

displacement, constant electric (superscript)	D
displacement constant, Wien's	b
displacement current	$\dot{D}$
displacement current	$\dot{D}/4\pi$
displacement, electric	D
displacement, electric [British]	D
displacement, electric [IUPAC]	D
displacement, electric [IUPAP]	D
displacement, electric [IUPAP]	D
displacement flux density	D
displacement flux, total electric	Ψ
displacement, linear	s
displacement, longitudinal	ξ
displacement of sound-bearing particle	ζ
displacement of sound-bearing particle	η
displacement of sound-bearing particle	ξ
displacement, particle	ξ
displacement, remanent or bias	D_0
displacement thickness	δ^*
displacement, transverse	η
displacement, volume	X
dissipated power per unit volume (piezoelectricity)	H
dissipation factor	D
dissipation factor [IUPAP]	δ
dissipative viscosity	η
dissociation, degree of	α
dissociation, degree of [IUPAC]	α
dissociation energy, nuclear	Λ
dissolution (subscript) [British]	d
distance	d
distance	l
distance	L
distance	s
distance, angular	θ
distance between adjacent principal foci of two lens units	Δ
distance between a principal plane of a lens system and the appropriate principal plane of a unit	p
distance between corresponding points of grating	d
distance beween image and corresponding principal plane	s'
distance between image and principal focus of image space	x'

distance between lens units in an optical system	d
distance between object and corresponding principal plane	s
distance between object and principal focus of object space	x
distance from focus to directrix of conic section	d
distance from source	r
distance, image	s'
distance, linear	s
distance, object	s
distance, radial	r
distance to extreme fiber from neutral axis	c
distribution function	f
distribution function	F
drag	D
drag [British]	D
drag (subscript)	D
drag, coefficient of	C_D
drag, coefficient of [British]	C_D
dry adiabatic processes (subscript)	a
dry air (subscript)	d
duration of rainfall	D
dynamic coefficient of eddy transfer	A
dynamic height	H_d
dynamic sensitivity of a phototube	s
dynamic viscosity	η
dynamic viscosity	μ
dynamic viscosity [British]	η
dynamic viscosity [British]	μ
east-west component of horizontal relative vorticity	ξ
eccentricity	e
eccentricity [British]	e
eddy transfer, dynamic coefficient of	A
effective current	I
effective molecular cross-section	σ
effective potential difference	V
effective wavelength	λ_e
efficiency	η
efficiency [British]	η
efficiency [IEC]	η
efficiency [IUPAP]	η
efficiency, luminous	K

efficiency, monochromatic luminous	K_λ
efficiency, voltage	η_v
Einstein temperature [IUPAP]	Θ_E
elastance	S
elastic compliance	s
elasticity, shear	μ
elasticity, shear modulus of	G
elasticity, shear modulus of	n
elasticity, volume modulus of	B
elastic stiffness	c
elastic stiffness constant	c_{33}
electric (subscript) [IUPAP]	e
electrical admittance	y_E
electrical admittance	Y_E
electrical capacitance	C_E
electrical conductance	g_E
electrical conductance	G_E
electrical energy [British]	W
electrical impedance, complex	z_E
electrical impedance, complex	Z_E
electrical power	P_E
electrical reactance	x_E
electrical reactance	X_E
electrical resistance	r_E
electrical resistance	R_E
electrical resistivity	ρ
electrical response	R_E
electrical susceptance	b_E
electrical susceptance	B_E
electric charge	q
electric charge	Q
electric charge [British]	Q
electric charge, quantity of [IEC]	Q
electric charge, volume density of	ρ
electric conductivity	σ
electric constant	Γ_e
electric current [British]	I
electric current [IUPAC]	i
electric current [IUPAC]	I
electric current [IUPAP]	I
electric current density	J
electric current density [IUPAC]	J

electric current density [IUPAP]	J
electric current density [IUPAP]	J
electric dipole moment [British]	μ
electric dipole moment [IUPAP]	p
electric dipole moment [IUPAP]	p
electric displacement	D
electric displacement [British]	D
electric displacement [IUPAC]	D
electric displacement [IUPAP]	D
electric displacement [IUPAP]	D
electric equivalent of heat	J
electric field [IUPAP]	E
electric field [IUPAP]	E
electric field, constant (superscript)	E
electric field strength	E
electric field strength [British]	E
electric field strength [IEC]	E
electric field strength [IEC]	K
electric field strength [IUPAC]	E
electric field, transverse	E_h
electric flux [British]	Ψ
electric flux [IEC]	Ψ
electric flux density	D
electric flux density [British]	D
electric flux density [IEC]	D
electric flux of induction	Ψ
electric flux, total	Ψ
electric force [British]	E
electric gradient	E
electric intensity	E
electric intensity [IEC]	E
electric intensity [IEC]	K
electricity, quantity of [IEC]	c
electric moment of atom	μ
electric moment of atom	μ_e
electric moment of dipole	μ
electric moment of dipole	μ_e
electric moment of molecule	μ
electric moment of molecule	μ_e
electric polarizability of a molecule [IUPAC]	α
electric polarizability of a molecule [IUPAC]	γ
electric polarization	P

DPMA

electric polarization [British]	P
electric potential	V
electric potential [British]	V
electric potential [IUPAC]	V
electric potential [IUPAP]	V
electric potential [IUPAP]	Φ
electric quadrupole moment	Q
electric susceptibility	η
electric susceptibility	χ_e
electric susceptibility [IUPAP]	χ_e
electrochemical equivalent	z
electrokinetic potential [British]	ζ
electrokinetic potential [IUPAC]	ζ
electrokinetic potential [British]	κ
electrolytic conductivity [IUPAC]	κ
electrolytic dissociation, degree of [IUPAC]	α
electromagnetic moment [IUPAP]	m
electromagnetic moment [IUPAP]	m
electromagnetic moment [IUPAP]	μ
electromagnetic moment [IUPAP]	μ
electromagnetic scalar potential	φ
electromechanical coupling factor	k
electromotive force	E
electromotive force	V
electromotive force [British]	E
electromotive force [British]	V
electromotive force [IEC]	E
electromotive force [IUPAC]	E
electron	e
electron [IUPAP]	e
electron affinity	φ
electron affinity, gross	φ_g
electronic Bohr magneton	μ_0
electronic charge	$-e$
electronic exit work function [British]	ϕ
electron mass	m_e
electron mass [IUPAP]	m
electron mass [IUPAP]	m_e
electron radius [IUPAP]	r_e
elementary charge [IUPAC]	e

DPMA

elementary charge [IUPAC]	e
ellipsoid flattening	f
elongation, total	δ
emissive power, monochromatic	J_λ
emissive power, monochromatic	J_ν
emissive power, total	J
emissivity	ϵ'
emissivity [British]	ϵ
emissivity, total [British]	ϵ
emissivity, total value of	ϵ_t'
emittance	ϵ
emittance, luminous	L
emittance, luminous [IUPAC]	H
emittance, luminous [IUPAP]	M
emittance, radiant	W
emittance, radiant [IUPAP]	M_e
empirical temperature [British]	t
empirical temperature [British]	θ
end correction for a tube	α
energy	E
energy [British]	E
energy [British]	W
energy [IEC]	E
energy [IEC]	U
energy [IEC]	W
energy [IUPAC]	E
energy [IUPAP]	E
energy [IUPAP]	U
energy, activation [British]	E
energy, average molecular kinetic	ϵ
energy balance in a nuclear reaction	Q
energy, constant (subscript)	E
energy density	E
energy, density of radiant	u
energy, electrical [British]	W
energy fluence of particles	F
energy, internal	U
energy, internal [British]	U
energy, kinetic	E_k
energy, kinetic	T
energy, kinetic [British]	T

DPMA

energy, kinetic [IUPAP]	E_k
energy, kinetic [IUPAP]	T
energy, luminous	Q
energy, molecular attraction [IUPAP]	ϵ
energy, nuclear dissociation	Λ
energy of gas, internal [British]	E
energy of gas, internal [British]	U
energy of gas, specific internal [British]	e
energy of gas, specific internal [British]	u
energy of vibration	E_v
energy per atom, internal	u
energy per atom, internal	u_m
energy per atom, intrinsic	u
energy per atom, intrinsic	u_m
energy per mole	E
energy per molecule, internal	u
energy per molecule, internal	u_m
energy per molecule, intrinsic	u
energy per molecule, intrinsic	u_m
energy per mole, internal	u
energy per mole, internal	U
energy per mole, internal	U_M
energy per mole, intrinsic	u
energy per mole, intrinsic	U
energy per mole, intrinsic	U_M
energy per unit mass, internal	u
energy per unit mass, intrinsic	u
energy, potential	E_p
energy, potential	U
energy, potential	V
energy, potential [British]	V
energy, potential [IUPAP]	E_p
energy, potential [IUPAP]	V
energy, radiant	U
energy required to form one ion pair [British]	W
energy, self	ϵ
energy, specific internal	u
energy, specific internal [British]	u
energy storage factor	Q
energy, strain [British]	U
energy, total	E
energy, total	U

DPMA

energy, total value of internal	U
energy, total value of intrinsic	U
enthalpy	H
enthalpy [British]	H
enthalpy [IUPAC]	H
enthalpy [IUPAP]	H
enthalpy, constant (subscript) [British]	h
enthalpy per atom	h_m
enthalpy per atom or molecule	h
enthalpy per mole	h
enthalpy per mole	H
enthalpy per mole	H_M
enthalpy per molecule	h_m
enthalpy per unit mass	h
enthalpy, specific	h
enthalpy, specific [British]	h
enthalpy, total	H
entropy	S
entropy [British]	S
entropy [IUPAC]	S
entropy [IUPAP]	S
entropy, constant (subscript)	S
entropy, constant (subscript) [British]	s
enthropy per atom	s
entropy per atom	s_m
entropy per mole	s
entropy per mole	S
entropy per mole	S_M
entropy per molecule	s
entropy per molecule	s_m
entropy per unit mass	s
entropy production rate	σ
entropy, specific	s
entropy, specific [British]	s
entropy, total	S
entropy, total [British]	S
epoch angle	ϵ
epoch angle	φ
equilibrium constant	K
equilibrium constant [IUPAC]	K
equilibrium constant [IUPAP]	K
equilibrium constant expressed in terms of activity [British]	K_a

equilibrium constant expressed in terms of concentration [British]	K_c
equilibrium constant expressed in terms of pressure [British]	K_p
equilibrium constant in terms of activity	K_a
equilibrium constant of reaction [British]	K
equivalent conductance of a negative ion [IUPAC]	Λ^-
equivalent conductance of a positive ion [IUPAC]	Λ^+
equivalent conductance of electrolyte [IUPAC]	Λ
equivalent conductance of ion [IUPAC]	Λ
equivalent conductivity	Λ
equivalent ionic conductance [British]	l
equivalent ionic conductance [British]	Λ
equivalent, Joule	J
equivalent noise conductance	G_n
equivalent quantities (subscript)	e
equivalents, number of	J
equivalent weight	M/z
evaporation (subscript) [British]	e
excess, mass [IUPAP]	Δ
excitation potential difference	V_e
expansion coefficient, cubic [IUPAP]	γ
expansion coefficient, linear [IUPAP]	α
expansivity, linear	α
expansivity, volume	β
exponential of x [IUPAP]	e^x
exposure	Σ
extent of chemical reaction	ξ
extent of chemical reaction [IUPAC]	ξ
extent of reaction [IUPAP]	ξ
external (subscript)	e
extinction (light) [IUPAC]	A
extinction (light) [IUPAC]	E
extinction coefficient	ϵ
extinction coefficient [IUPAC]	κ
extinction coefficient [IUPAP]	κ
extinction coefficient, molar [IUPAC]	ϵ
extinction coefficient, molecular [British]	ϵ
extrapolated value of a length (subscript) [British]	e
extraterrestrial solar radiant intensity	I_0
extraterrestrial solar radiant intensity	J_0
extreme fiber from neutral axis, distance to	c

factor	c
factor	C
factor	k
factor [British]	F
factor, normalization	N
factor of safety	N
Faraday constant	F
Faraday constant [British]	F
Faraday constant [IUPAC]	F
Faraday constant [IUPAC]	F
Faraday constant [IUPAP]	F
field strength, electric	E
field strength, electric [British]	E
field strength, electric [IEC]	E
field strength, electric [IEC]	K
field strength, electric [IUPAC]	E
field strength, magnetic	H
field strength, magnetic	$\mathbf{H}$
field strength, magnetic [British]	H
field strength, magnetic [IUPAC]	H
figure of merit of inductance	Q
filament (subscript)	f
filament supply voltage, alternating-current	E_f
final values (subscript)	f
fine structure constant	α
fine structure constant [IUPAP]	α
finite increase of x [IUPAP]	Δx
fission rate	F
flare coefficient in a horn	m
flattening, ellipsoid	f
flow rate, heat	$\dot{Q}$
flow rate, heat [British]	$\dot{Q}$
flow rate, mass	$\dot{m}$
flow rate, mass [British]	$\dot{m}$
flow rate, volume [British]	Q
flow rate, volume [British]	$\dot{V}$
flow rate, volumetric	Q
fluidity	φ
fluidity [British]	ϕ
fluidity [IUPAC]	φ
flux density	B
flux density, electric	D
flux density, electric [British]	D

flux density, electric [IEC]	D
flux density, magnetic	$\mathbf{B}$
flux density, magnetic [British]	$\mathbf{B}$
flux density, magnetic [IEC]	B
flux density, propagation	Ξ
flux density, radiant	$\mathcal{E}$
flux density, radiant	$\mathcal{R}$
flux density, radiant	W
flux, electric [British]	Ψ
flux, electric [IEC]	Ψ
flux, light [IUPAC]	Φ
flux, luminous	F
flux, luminous [British]	F
flux, luminous [British]	Φ
flux, luminous [IUPAP]	Φ
flux, magnetic	Φ
flux, magnetic [British]	Φ
flux, magnetic [IEC]	Φ
flux, magnetic [IUPAP]	Φ
flux of induction, electric	Ψ
flux, radiant	P
flux, radiant	Φ
flux, radiant [IUPAP]	P
flux, radiant [IUPAP]	Φ_e
flux, total dielectric	Ψ
flux, total electric	Ψ
focal length of image space	f'
focal length of object space	f
focus of conic section	f
force	F
force [British]	$\mathbf{F}$
force [IEC]	F
force [IUPAC]	F
force [IUPAP]	F
force [IUPAP]	$\mathbf{F}$
force, concentrated	F
force, concentrated	P
force, concentrated	Q
force constant	k
force due to gravity [IEC]	G
force due to gravity [IUPAC]	G
force due to gravity [IUPAC]	W

force, electric [British]	E
force, electromotive	E
force, electromotive	V
force, gravitational	F
force, gravitational	F_g
force, horizontal	F_h
force in membrane	T
force in membrane, stretching	F
force in string	T
force in string, stretching	F
force, magnetizing	H
force, magnetizing [British]	H
force, magnetomotive	$\mathcal{F}$
force, moment of	M
force, moment of [IUPAC]	M
force, resultant	F_r
force, shear [British]	Q
force, vertical	F_v
formality	F
fourth virial coefficient	D
fraction of incident radiant power which is absorbed [IUPAC]	α
fraction of incident radiant power which is reflected [IUPAC]	ρ
fraction of incident radiant power which is transmitted [IUPAC]	τ
free energy, Gibbs function	F
free energy, Gibbs function	G
free energy, Helmholtz function	A
free energy, Helmholtz function [British]	F
free energy, specific [British]	f
free-fall acceleration [IUPAC]	g
free path	l
free path	λ
free path, mean	$\bar{l}$
free path, mean	$\bar{\lambda}$
free path, Tait	l_T
free path, Tait	λ_T
free space, permeability of [British]	μ_0
free space, permeability of [IEC]	μ_0
free space, permittivity of	ϵ_0
free space, permittivity of [British]	ϵ_0

free space, permittivity of [IEC]	ε_0
free space, permittivity of [IEC]	ϵ_0
free space, reciprocal permeability of	μ_0
frequency	f
frequency	ν
frequency [British]	f
frequency [British]	ν
frequency [IEC]	f
frequency [IEC]	ν
frequency [ISO]	f
frequency [ISO]	ν
frequency [IUPAC]	f
frequency [IUPAC]	ν
frequency [IUPAP]	f
frequency [IUPAP]	ν
frequency (subscript)	f
frequency, angular	ω
frequency, angular [British]	ω
frequency, angular [IEC]	ω
frequency, angular [ISO]	ω
frequency, angular [IUPAC]	ω
frequency, angular [IUPAP]	ω
frequency, circular	ω
frequency constant	N
frequency deviation	δ
frequency, maximum	ν_{max}
frequency, natural	f_r
frequency of free vibration with damping, angular	ω'
frequency of molecular collision	Z
fequency, photoelectric threshold	ν_0
frequency, radiation	ν
frequency, rotational	n
frequency, rotational [British]	n
frequency, rotational [IEC]	n
frequency, rotational [ISO]	n
frequency without damping, angular	ω
frictional viscosity	η
friction, angle of [British]	ϕ
friction coefficient [IUPAC]	f
friction coefficient [IUPAP]	f
friction, coefficient of	μ
friction, coefficient of [British]	μ

friction, coefficient of kinetic	μ
friction, coefficient of kinetic	μ_k
friction, coefficient of rolling	μ_r
friction, coefficient of sliding	μ
friction, coefficient of sliding	μ_k
friction, coefficient of starting	μ_s
friction, coefficient of static	μ_s
friction factor	f
friction factor [British]	f
friction head	h_f
friction head [British]	h_f
friction velocity	u^*
friction velocity [British]	U_τ
friction velocity [British]	v^*
frost point (subscript)	f
Froude number	F
Froude number [British]	Fr
Froude number (fluid mechanics) [British]	F
fugacity	f
function	f
function, distribution	f
function, distribution	F
function, Lagrangian	L
function of [British]	f
function of x [IUPAP]	$f(x)$
fusion (subscript) [British]	f
fusion processes (subscript)	f
gain of amplifier, power	A
gain of amplifier, power	A_p
gain of amplifier, voltage	A
gain of amplifier, voltage	A_v
galvanometer deflection	δ
gamma function	Γ
gamma function [British]	Γ
gamma-ray constant, specific	Γ
gas (subscript)	G
gas (subscript) [British]	G
gas (subscript) [IUPAP]	g
gas amplification factor	μ
gas constant	R
gas constant [IUPAC]	R
gas constant [IUPAC]	$\mathbf{R}$
gas constant, characteristic [British]	R

gas constant, characteristic [British]	$\mathscr{R}$
gas constant for gas of molecular weight M [British]	R_M
gas constant, molar	R_o
gas constant, molar [British]	R_o
gas constant, molar [British]	R
gas constant, molecular	k
gas constant per mole [IUPAP]	R
gas constant, specific	r
gas constant, specific	R
gas constant, specific [British]	$\mathscr{R}$
gas constant, specific [British]	R
gas constant, universal	R
gas constant, universal	R^*
gas constant, universal [British]	R_o
gas constant, universal [British]	R
generalized coordinate	q
generalized coordinate [IUPAP]	q
generalized momentum	p
generalized momentum [IUPAP]	p
geopotential	ϕ
geopotential altitude	H
geostrophic wind (subscript)	g
Gibbs function	G
Gibbs function [British]	G
Gibbs function [IUPAC]	F
Gibbs function [IUPAC]	G
Gibbs function [IUPAP]	G
Gibbs function equation, integration constant of	I
Gibbs function, partial molal	μ
Gibbs function per atom	g_m
Gibbs function per mole	g
Gibbs function per mole	G
Gibbs function per mole	G_M
Gibbs function per molecule	g_m
Gibbs function per unit mass	g
Gibbs function, specific	g
Gibbs function, specific [British]	g
Gibbs function, total value	G
gradient wind (subscript)	gr
Graetz number [British]	Gz
Grashof number	Gr

Grashof number [British]	Gr
grating lines, total number of	N
grating space	d
gravitational acceleration	g
gravitational acceleration [British]	g
gravitational acceleration [ISO]	g
gravitational acceleration [IUPAP]	g
gravitational acceleration, standard [IUPAP]	g_n
gravitational constant, Newtonian	G
gravitational force	F
gravitational force	F_g
gravitational force [IUPAC]	G
gravity, local acceleration due to	g_L
gravity, standard acceleration due to	g_o
gravity, standard acceleration due to [British]	g_n
grid admittance with plate load	y_g
grid conductance	g_g
grid conductance at zero frequency and constant plate potential	k_g
grid control ratio	μ
grid-plate transadmittance	y_{pg}
grid-plate transconductance	g_m
grid-plate transconductance	g_{pg}
grid resistance	R_g
grid, screen (subscript)	s
grid, suppressor (subscript)	su
gross electron affinity	φ_g
gross work function per unit charge	φ_g
group (subscript)	g
group velocity	u_g
group velocity	v_g
group velocity [IUPAP]	c_g
growth, period of [British]	τ
gyromagnetic ratio	g
gyromagnetic ratio [IUPAP]	γ
half-angle subtended at point object by objective of microscope	α
half-life [IUPAP]	$T_{1/2}$
half-life (radioactivity)	T
half-life of radioisotope [British]	T
half-life of radioisotope [British]	$T_{1/2}$
Hall field	E_h

Hamiltonian function	H
Hamiltonian function [IUPAP]	H
Hamiltonian function, perturbing	$\mathcal{H}$
Hamiltonian operator	H
head	h
head, friction	h_f
head, friction [British]	h_f
head, maximum	h_{max}
head, maximum [ASA]	h_{max}
head, potential [British]	z
head, pressure	h_p
head, pressure [British]	h_p
head, velocity	h_v
head, velocity [British]	h_v
heat [IUPAC]	q
heat [IUPAC]	Q
heat (subscript)	h
heat (subscript)	q
heat capacity [IUPAC]	C
heat capacity at constant pressure	C_p
heat capacity at constant pressure [British]	C_p
heat capacity at constant pressure, specific	c_p
heat capacity at constant pressure, specific [British]	c_p
heat capacity at constant volume	C_v
heat capacity at constant volume [British]	C_v
heat capacity at constant volume, specific	c_v
heat capacity at constant volume, specific [British]	c_v
heat capacity, molar [IUPAP]	C_p
heat capacity, molar [IUPAP]	C_v
heat capacity per atom	c_m
heat capacity per mole	c
heat capacity per mole	C
heat capacity per mole	C_M
heat capacity per mole [British]	C
heat capacity per molecule	c_m
heat capacity per unit mass	c
heat capacity, specific	c
heat capacity, specific [British]	c
heat capacity, specific [IUPAP]	c_p
heat capacity, specific [IUPAP]	c_v
heat capacity, total value of	C
heat content	H

DPMA

heat content, relative	L
heat, electric equivalent of	J
heat entering system	q
heat equivalent of work	A
heater (subscript)	h
heat flow path, length of	l
heat flow path, length of	L
heat flow rate	q
heat flow rate	$\dot{Q}$
heat flux intensity [British]	q
heat flux intensity [British]	ϕ
heat, mechanical equivalent of	J
heat of reaction or phase change per atom or molecule	Δh
heat of reaction or phase change per atom or molecule	Δh_m
heat of reaction or phase change per mole	Δh
heat of reaction or phase change per mole	ΔH
heat of reaction or phase change per mole	ΔH_M
heat of reaction or phase change per unit mass	Δh
heat of reaction or phase change, total value	ΔH
heat, quantity of	Q
heat, quantity of [British]	Q
heat, sensible [British]	q_s
heat. sensible [British]	Q_s
heat, total [British]	H
heat transfer coefficient	h
heat transfer, coefficient of [British]	α
heat transfer coefficient, over-all	U
heat transfer coefficient, over-all [British]	U
heat transfer coefficient, surface	h
heat transfer factor	j
heat transfer, rate of	$\dot{Q}$
height	h
height	y
height [British]	h
height [IEC]	h
height [ISO]	h
height [IUPAC]	h
height, dynamic	H_d
height of homogeneous atmosphere	H
height of image	y'

Helmholtz free energy [IUPAC]	A
Helmholtz free energy [IUPAC]	F
Helmholtz function	A
Helmholtz function [IUPAP]	F
Helmholtz function per atom or molecule	a_m
Helmholtz function per mole	A_M
Henry law constant	H
Hermitian conjugate of A [IUPAP]	$A^\dagger$
Hertzian vector	Π
higher (subscript)	H
homogeneous atmosphere, height of	H
horizon distance	ϵ
horizontal (subscript)	h
horizontal axis	X
horizontal force	F_h
horizontal surface (subscript)	h
horizontal thrust	H
humidity	H
humidity [British]	H
humidity, absolute	ρ
humidity at adiabatic saturation temperature	H_a
humidity at adiabatic saturation temperature [British]	H_a
humidity at saturation	H_s
humidity at saturation [British]	H_s
humidity at wet-bulb temperature	H_w
humidity at wet-bulb temperature [British]	H_w
humidity, relative	f
humidity, relative	r
humidity, relative	U
humidity, relative [British]	H_R
humidity, relative [British]	ϕ
humidity, specific	q
hydrolysis constant	K_h
hydrolysis, degree of	h
hydrostatic piezoelectric strain constant	d_h
hyperfine quantum number [IUPAP]	F
ice (subscript)	i
illuminance	E
illumination [IUPAC]	E
illumination [IUPAP]	E
illumination, amount of	E

DPMA

image distance	s'
image, height of	y'
image, length of	y'
imaginary part of z [IUPAP]	z''
imaginary unit	i
imaginary unit	j
imaginary unit [IUPAP]	i
imaginary unit [IUPAP]	j
impact parameter	p
impact parameter [IUPAP]	b
impedance	z
impedance	Z
impedance [British]	Z
impedance [IEC]	Z
impedance [IUPAC]	Z
impedance [IUPAP]	Z
impedance, characteristic	Z_0
impedance, complex acoustical	z_A
impedance, complex acoustical	Z_A
impedance, complex electrical	z_E
impedance, complex electrical	Z_E
impedance, complex mechanical	z_M
impedance, complex mechanical	Z_M
impedance, complex rotational	z_R
impedance, complex rotational	Z_R
impedance, input	Z_i
impedance, load	Z_L
impedance, mutual	Z_m
impedance, output	Z_o
impedance, reciprocal	Y
impedance, total	Z_T
impulse	I
incidence, angle of	i
incidence, angle of	φ
incidence, principal angle of	$\bar{\varphi}$
incident radiant power which is absorbed, fraction of [IUPAC]	α
incident radiant power which is reflected, fraction of [IUPAC]	ρ
incident radiant power which is transmitted, fraction of [IUPAC]	τ
inclination of orbit	ι

increment [IEC]	Δ
increment, finite	δ
increment, finite	Δ
index, absorption	κ
index of refraction	n
index of refraction [British]	n
index of refraction [IUPAC]	n
indicated (subscript)	i
inductance	L
inductance [British]	L
inductance [IUPAP]	L
inductance, mutual	L_{12}
inductance, mutual	M
inductance, mutual [British]	L_{mn}
inductance, mutual [British]	M
inductance, mutual [IUPAC]	L_{12}
inductance, mutual [IUPAC]	M
inductance, reciprocal	Γ
induction coefficient	c
induction density, electric	D
induction density, magnetic	B
induction, electric flux of	Ψ
induction, magnetic	B
induction, magnetic [British]	B
induction, magnetic [IEC]	B
induction, magnetic [IUPAC]	B
induction, magnetic [IUPAP]	B
induction, magnetic [IUPAP]	B
inductive capacity, specific	ϵ
inductive capacity, specific	ϵ_r
inductive current	I_L
inductive reactance	X_L
inductivity	μ
inertance	M
inertia of photographic plate	i
inertia proportionality factor	g_c
initial number of atoms	N_0
initial number of nuclei	N_0
initial speed	u_o
initial speed	v_o
initial value (subscript)	i
initial value (subscript)	o
initial velocity	u_o

initial velocity .... v_o
inner potential .... V_i
input (subscript) i
input admittance with plate load .... y_g
input conductance .... g_g
input impedance .... Z_i
input power .... P_i
inside (subscript) i
instantaneous current i
instantaneous current through resistance .... i_R
instantaneous density of the medium .... ρ
instantaneous particle velocity .... u_i
instantaneous sound pressure .... p_i
integration constant of Gibbs function equation ... I
intensity I
intensity, acoustic I
intensity, electric .... $\mathbf{E}$
intensity, electric [IEC] .... E
intensity, electric [IEC] .... K
intensity, heat flux [British] .... q
intensity, heat flux [British] .... ϕ
intensity, luminous I
intensity, luminous [British] I
intensity, luminous [IUPAC] I
intensity, luminous [IUPAP] I
intensity, magnetic .... $\mathbf{H}$
intensity, magnetic .... $\mathbf{H}$
intensity, magnetic [IEC] .... H
intensity of magnetization [British] I
intensity of magnetization [British] .... M
intensity of radiation I
intensity of radiation [British] I
intensity, radiant .... J
intensity, radiant [IUPAP] .... I_e
interaction energy between molecules i and j
    [IUPAP] .... V_{ij}
interaction energy between molecules i and j
    [IUPAP] .... φ_{ij}
interfacial area per common volume [British] ..... a
internal (subscript) i
internal conversion coefficient [IUPAP] .... α
internal energy .... U
internal energy [British] .... U

DPMA

internal energy [IUPAP]	U
internal energy of gas [British]	E
internal energy of gas [British]	U
internal energy of gas, specific [British]	e
internal energy of gas, specific [British]	u
internal energy, specific [British]	u
internal energy, total [British]	U
internal energy, total value	U
interplanar distance (Bragg law)	d
interval, time [IUPAC]	τ
intrinsic angular momentum	s
intrinsic energy, total value	U
intrinsic induction	B_i
inverse transconductance	g_{sp}
inverse transconductance	g_n
ion charge number [IUPAC]	z
ionic conductance, equivalent [British]	l
ionic conductance, equivalent [British]	Λ
ionic strength [British]	I
ionic strength [IUPAC]	I
ionic strength [IUPAC]	μ
ionic strength [IUPAP]	I
ionization, degree of	α
ionization potential	V_i
irradiance	$\mathcal{E}$
irradiance	H
irradiance	W
irradiance [IUPAP]	E_e
irradiance, spectral	H_λ
isobaric processes (subscript)	p
isothermal processes (subscript)	θ
isotopic number [British]	I
isotopic spin [British]	T
Joule equivalent	J
Joule-Thomson coefficient	μ
Joule-Thomson coefficient [British]	μ
Joule-Thomson coefficient [IUPAC]	μ
Joule-Thomson coefficient [IUPAP]	μ
Kelvin temperature	T
Kerr constant	K
kinematic viscosity	ν
kinematic viscosity [British]	ν
kinematic viscosity [IUPAC]	ν

DPMA

kinematic viscosity [IUPAP]	ν
kinematic viscosity, coefficient of	ν
kinetic (subscript)	k
kinetic (subscript) [IUPAP]	k
kinetic energy	E_k
kinetic energy	T
kinetic energy [British]	T
kinetic energy [IUPAP]	E_k
kinetic energy [IUPAP]	T
kinetic energy, average molecular	ϵ
kinetic friction, coefficient of	μ_k
kinetic potential	L
kinetic pressure	q
Knudsen number [British]	Kn
Lagrangian function	L
Lagrangian function [IUPAP]	L
lambda particle	Λ°
Landé factor	g
lapse rate of temperature, actual	γ
lapse rate of temperature, adiabatic	Γ
Larmor (angular) frequency [IUPAP]	ω_L
latent heat of phase change	L
latent heat, specific [British]	l
latitude	ϕ
latitudinal rate of variation of Coriolis parameter	β
leakage coefficient [IEC]	σ
leakage coefficient, magnetic	σ
left (subscript)	L
length	l
length	L
length	s
length [British]	l
length [IEC]	l
length [ISO]	l
length [IUPAC]	l
length [IUPAP]	l
length, mixing	l
length of arc	s
length of image	y'
length of path	s
length of path [ISO]	s
length of prism base	t

length of vibrating string, rod, or tube ... l
length, optical ... Δ
length, rest ... l_0
lens system, power of ... D
lens zone, radius of ... h
lethargy [British] ... u
level width [British] ... Γ
level width [IUPAP] ... Γ
lifetime, neutron [British] ... l
light flux [IUPAC] ... Φ
light, quantity of ... Q
light, quantity of [IUPAC] ... Q
linear absorption coefficient ... μ
linear absorption coefficient [British] ... μ
linear absorption coefficient [IUPAP] ... μ
linear absorption coefficient [IUPAP] ... μ_l
linear acceleration ... $\mathbf{a}$
linear acceleration [British] ... a
linear acceleration [British] ... f
linear acceleration [IEC] ... a
linear charge density ... λ
linear coefficient of thermal expansion [British] ... α
linear current density ... A
linear current density [IEC] ... A
linear current density [IEC] ... α
linear density ... λ
linear density ... ρ
linear displacement ... $\mathbf{s}$
linear distance ... s
linear energy transfer ... L
linear expansion coefficient ... α
linear expansion coefficient [IUPAP] ... α
linear expansivity ... α
linear range [IUPAP] ... R
linear range [IUPAP] ... R_l
linear stopping power [IUPAP] ... S
linear stopping power [IUPAP] ... S_l
linear velocity ... $\mathbf{u}$
linear velocity ... $\mathbf{v}$
linear velocity [British] ... u
linear velocity [British] ... v
linear velocity [British] ... w

linear velocity [IEC]	v
lines of a grating, total number of	N
liquid (subscript)	L
liquid (subscript) [British]	L
liquid water content per unit volume	ρ_e
load, concentrated	F
load, concentrated	P
load, concentrated	Q
load factor [British]	N
load factor [British]	Ω
load impedance	Z_L
load moment of inertia	J_L
load per unit displacement	k
load resistance	R_L
load, total	P
load, total	W
local acceleration due to gravity	g_L
local surface friction coefficient [British]	c_f
logarithmic decrement [ISO]	Λ
longitude	λ
longitudinal coupling factor (piezoelectricity)	k_{33}
longitudinal displacement	ξ
Lorentz unit	L
Loschmidt number	n_0
loss angle [IUPAP]	δ
loudness level [IUPAP]	L_N
loudness level [IUPAP]	Λ
luminance	B
luminance [British]	L
luminance [IUPAC]	B
luminance [IUPAC]	L
luminance [IUPAP]	L
luminance factor [British]	β
luminosity	K_λ
luminosity factor	K
luminous efficiency	K
luminous efficiency, monochromatic	K_λ
luminous emittance	L
luminous emittance [IUPAC]	H
luminous emittance [IUPAP]	M
luminous energy	Q
luminous flux	F

luminous flux [British]	F
luminous flux [British]	Φ
luminous flux [IUPAP]	Φ
luminous intensity	I
luminous intensity [British]	I
luminous intensity [IUPAC]	I
luminous intensity [IUPAP]	I
Mach angle	μ
Mach angle [British]	μ
Mach number	M
Mach number [British]	M
macroscopic cross-section [British]	Σ
macroscopic cross-section [IUPAP]	Σ
magnetic constant	Γ_m
magnetic dipole moment [British]	μ
magnetic dipole moment [IUPAP]	j
magnetic dipole moment [IUPAP]	j
magnetic field [IUPAP]	H
magnetic field [IUPAP]	H
magnetic field strength	H
magnetic field strength	H
magnetic field strength [British]	H
magnetic field strength [IEC]	H
magnetic field strength [IUPAC]	H
magnetic flux	Φ
magnetic flux [British]	Φ
magnetic flux [IEC]	Φ
magnetic flux [IUPAP]	Φ
magnetic flux density [British]	B
magnetic flux density [IEC]	B
magnetic induction	B
magnetic induction [British]	B
magnetic induction [IEC]	B
magnetic induction [IUPAC]	B
magnetic induction [IUPAP]	B
magnetic induction [IUPAP]	B
magnetic intensity	H
magnetic intensity	H
magnetic intensity [IEC]	H
magnetic leakage coefficient	σ
magnetic moment	m
magnetic moment	μ

magnetic moment of atom	μ
magnetic moment of atom	μ_m
magnetic moment of dipole	μ
magnetic moment of dipole	μ_m
magnetic moment of electron [IUPAP]	μ_e
magnetic moment of molecule	μ
magnetic moment of molecule	μ_m
magnetic moment of neutron [IUPAP]	μ_n
magnetic moment of particle [IUPAP]	μ
magnetic moment of proton [IUPAP]	μ_p
magnetic permeability	μ
magnetic permeability [IUPAC]	μ
magnetic permeability, relative	μ
magnetic permeability, relative	μ_r
magnetic polarization	$\mathbf{M}$
magnetic polarization [British]	I
magnetic polarization [British]	M
magnetic polarization [IUPAP]	J
magnetic poles, number of pairs of	p
magnetic pole strength	m
magnetic pole strength	p
magnetic potential	$\mathcal{F}$
magnetic pressure number	S
magnetic quantum number [IUPAP]	m_i
magnetic quantum number [IUPAP]	M
magnetic Reynolds number	R_M
magnetic rotation, specific	V
magnetic rotation, specific	ω
magnetic scalar potential	$\mathcal{F}$
magnetic shell strength	I
magnetic susceptibility	k
magnetic susceptibility	χ_m
magnetic susceptibility [IUPAC]	χ
magnetic susceptibility [IUPAP]	χ_m
magnetic susceptibility, mass [British]	χ
magnetic susceptibility, specific	χ
magnetic susceptibility, volume [British]	κ
magnetic vector potential	$\mathbf{A}$
magnetic vector potential [British]	A
magnetization	$\mathbf{M}$
magnetization [IEC]	J
magnetization [IUPAC]	M

magnetization [IUPAP]	M
magnetization [IUPAP]	$\mathbf{M}$
magnetization, intensity of [British]	I
magnetization, intensity of [British]	M
magnetization, specific	σ
magnetizing force	H
magnetizing force	$\mathbf{H}$
magnetizing force [British]	H
magnetomotive force	$\mathcal{F}$
magnetomotive force [British]	F
magnetomotive force [British]	M
magnetomotive force [IEC]	F
magnetomotive force [IEC]	F_m
magnetomotive force [IEC]	$\mathcal{F}$
magneton, Bohr	μ_0
magneton, nuclear	μ_I
magneton, orbital	μ_0
magnetostriction constant	K
magnification, angular	γ
magnification, linear	m
major hydraulic area	A
mass	m
mass [British]	m
mass [IEC]	m
mass [IUPAC]	m
mass [IUPAP]	m
mass (subscript)	m
mass absorption coefficient [IUPAP]	μ_m
mass, atomic [IUPAP]	M_a
mass concentration	c
mass density	ρ
mass density [British]	ρ
mass diffusion coefficient	D
mass diffusivity	D
mass, electron	m_e
mass, electron [IUPAP]	m
mass, electron [IUPAP]	m_e
mass excess [IUPAP]	Δ
mass flow rate	$\dot{m}$
mass flow rate [British]	$\dot{m}$
mass fraction	w
mass fraction of substance B [IUPAP]	w_B

mass magnetic susceptibility [British]	χ
mass, meson [IUPAP]	m_μ
mass, meson [IUPAP]	m_π
mass, molar [IUPAC]	M
mass, neutron [IUPAP]	m_n
mass, nuclear [IUPAP]	M
mass, nuclear [IUPAP]	M_N
mass number	A
mass number [IUPAP]	A
mass of atom	m_m
mass of electron	m
mass of electron [British]	m
mass of molecule	m_m
mass per unit area	ρ
mass per unit area	σ
mass per unit length	λ
mass per unit length	ρ
mass per unit volume	D
mass per unit volume	ρ
mass, proton [IUPAP]	m_p
mass, reduced	μ
mass, reduced [IUPAP]	μ
mass, rest	m_0
mass velocity	G
mass velocity [British]	G
maximum (subscript)	m
maximum (subscript)	max
maximum current	I_m
maximum frequency	ν_{max}
maximum head	h_{max}
maximum head [ASA]	h_{max}
maximum isothermal work function	A
maximum isothermal work function per atom or molecule	a_m
maximum particle velocity	u_m
maximum peak current	I_m
maximum peak current	I_{mp}
maximum peak potential difference	V_m
maximum peak potential difference	V_{mp}
maximum potential difference	V_m
maximum sound pressure	p_m
mean (subscript)	m

DPMA

mean free path	l
mean free path	$\bar{\lambda}$
mean free path [British]	l
mean free path [British]	λ
mean free path [IUPAC]	l
mean free path [IUPAP]	l
mean-life	τ
mean-life [IUPAP]	τ
mean-life of radioisotope [British]	t_m
mean value (subscript) [British]	m
mechanical admittance	y_M
mechanical admittance	Y_M
mechanical compliance	C_M
mechanical conductance	g_M
mechanical conductance	G_M
mechanical equivalent of heat	J
mechanical impedance, complex	z_M
mechanical impedance, complex	Z_M
mechanical power	P_M
mechanical reactance	x_M
mechanical reactance	X_M
mechanical resistance	r_M
mechanical resistance	R_M
mechanical response	R_M
mechanical susceptance	b_M
mechanical susceptance	B_M
meson mass [IUPAP]	m_μ
meson mass [IUPAP]	m_π
micron	μ
minimum (subscript)	min
minimum nadir angle	η_0
mixing length	l
mixing ratio	w
modulation factor	m
modulus, decay	τ
modulus of compression [IUPAC]	K
modulus of elasticity	E
modulus of elasticity	Y
modulus of elasticity [British]	E
modulus of elasticity [IUPAC]	E
modulus of elasticity [IUPAP]	E
modulus of elasticity, bulk	K

modulus of elasticity, shear	G
modulus of elasticity, shear	n
modulus of elasticity, volume	B
modulus of rigidity [British]	G
modulus of rupture	R
modulus of rupture [British]	R
modulus of section	Z
modulus of section [British]	Z
modulus of z [IUPAP]	$|z|$
modulus, piezoelectric strain	δ
modulus, shear [British]	G
molality	m
molality [British]	m
molality [IUPAC]	m
molality of solution [IUPAP]	m
molar absorptivity [IUPAC]	ϵ
molar concentration of substance B	c_B
molar concentration of substance B [IUPAC]	$[B]$
molar concentration of substance B [IUPAC]	c_B
molar concentration of substance B [IUPAC]	$c(B)$
molar concentration of substance B [IUPAP]	c_B
molar concentration of substance X [British]	c_x
molar concentration of substance X [British]	C_x
molar concentration of substance X [British]	$[X]$
molar conductance of electrolyte [IUPAC]	Λ
molar conductance of ion [IUPAC]	Λ
molar decadic absorption [IUPAC]	ϵ
molar extinction coefficient [IUPAC]	ϵ
molar heat capacity [IUPAP]	C_p
molar heat capacity [IUPAP]	C_v
molar mass	M
molar mass [IUPAC]	M
molar mass of substance B [IUPAP]	M_B
molecular attraction energy [IUPAP]	ϵ
molecular collision frequency	Z
molecular concentration	n
molecular concentration [IUPAC]	C
molecular concentration at zero degree centigrade and one atmosphere	n_0
molecular conductivity	μ
molecular cross-section, effective	σ
molecular density	n

DPMA

molecular extinction coefficient [British]	ϵ
molecular gas constant	k
molecular kinetic energy, average	ϵ
molecular mass	m
molecular mass	$\bar{m}$
molecular mass [IUPAC]	m
molecular mass [IUPAP]	m
molecular momentum vector [IUPAP]	p
molecular position vector [IUPAP]	r
molecular-scale temperature geometric gradient	L_M
molecular-scale temperature geopotential gradient	L'_M
molecular velocity vector [IUPAP]	c
molecular velocity vector [IUPAP]	u
molecular viscosity, coefficient of	μ
molecular volume	V
molecular weight	M
molecular weight [British]	M
molecule, collision diameter of	σ
molecule, diameter of [IUPAC]	D
molecule, diameter of [IUPAC]	σ
molecules in a chemical reaction, stoichiometric number of	ν
molecules, number of	N
molecules, number of [British]	N
molecules, number of [IUPAC]	N
molecules, stoichiometric number of [British]	ν
molecules, stoichiometric number of [IUPAC]	ν
mole factor	i
mole fraction	x
mole fraction [British]	x
mole fraction [IUPAC]	x
mole fraction [IUPAC]	X
mole fraction [IUPAC]	y
mole fraction basis (subscript)	x
mole fraction, gas phase	y
mole fraction of substance B [IUPAP]	x_B
mole fraction of substance B [IUPAP]	X_B
mole ratio of solution [IUPAP]	r
moles, number of	n
moles, number of [British]	n
moles, number of [IUPAC]	n
moment	M

moment [British]	M
moment, magnetic	$\mathbf{m}$
moment, magnetic	μ
moment of area	Q
moment of atom, electric	μ
moment of atom, electric	μ_e
moment of atom, magnetic	μ
moment of atom, magnetic	μ_m
moment of dipole, electric	μ
moment of dipole, electric	μ_e
moment of dipole, magnetic	μ
moment of dipole, magnetic	μ_m
moment of electron, magnetic [IUPAP]	μ_e
moment of force	M
moment of force [IEC]	M
moment of force [IUPAC]	M
moment of force [IUPAP]	M
moment of force [IUPAP]	$\boldsymbol{M}$
moment of inertia	I
moment of inertia	I_{xx}
moment of inertia	J
moment of inertia [British]	I
moment of inertia [IEC]	I
moment of inertia [IEC]	J
moment of inertia [IUPAC]	I
moment of inertia [IUPAP]	I
moment of inertia [IUPAP]	J
moment of inertia, areal	I
moment of inertia, load	J_L
moment of inertia, polar	J
moment of inertia, rectangular	I
moment of molecule, electric	μ
moment of molecule, electric	μ_e
moment of molecule, magnetic	μ
moment of molecule, magnetic	μ_m
moment of neutron, magnetic [IUPAP]	μ_n
moment of particle, magnetic [IUPAP]	μ
moment of proton, magnetic [IUPAP]	μ_p
momentum	M
momentum [British]	p
momentum [IUPAP]	p
momentum [IUPAP]	$\boldsymbol{p}$

momentum (subscript)	M
momentum, generalized	p
momentum thickness of boundary layer	θ
monochromatic (subscript)	λ
monochromatic luminous efficiency	K_λ
most probable speed [IUPAP]	$\bar{c}$
most probable speed [IUPAP]	$\hat{u}$
motional capacitance constant	Γ
muon	μ
muon [IUPAP]	μ
mutual conductance	G_M
mutual conductance [British]	g_m
mutual impedance	Z_m
mutual inductance	L_{12}
mutual inductance	M
mutual inductance [British]	L_{mn}
mutual inductance [British]	M
mutual inductance [IEC]	L_{mn}
mutual inductance [IEC]	M
mutual inductance [IUPAC]	L_{12}
mutual inductance [IUPAC]	M
mutual inductance [IUPAP]	L_{12}
mutual inductance [IUPAP]	M
nadir angle	η
nadir angle, minimum	η_0
nadir angle, object	η^*
Naperian logarithms, base of	e
Naperian logarithms, base of	ϵ
natural coordinates	s, n, z
natural frequency	f_r
net work function per unit charge	φ
neutral axis, distance to extreme fiber from	c
neutral charge (superscript) [IUPAP]	0
neutrino	ν
neutrino [IUPAP]	ν
neutron	n
neutron [IUPAP]	n
neutron density [British]	n
neutron density, thermal [British]	n
neutron mass [IUPAP]	m_n
neutron number [IUPAP]	N
neutrons in nucleus, number of [British]	N

neutrons, number of [British]	n
normal (subscript)	n
normal (subscript) [IUPAP]	n
normal angle	θ
normal component (subscript)	n
normality	C
normalization factor	N
normal strain	ϵ
normal stress	s
normal stress	σ
normal stress [British]	f
normal stress [British]	σ
normal stress [IUPAP]	σ
north-south component of horizontal relative vorticity	η
nuclear Bohr magneton	μ_I
nuclear charge	Z
nuclear dissociation energy	Λ
nuclear magneton	μ_I
nuclear magneton [IUPAP]	μ_N
nuclear mass [IUPAP]	M
nuclear mass [IUPAP]	M_N
nuclear radius	r
nuclear radius [British]	r
nuclear radius [British]	R
nuclear radius [IUPAP]	R
nuclear spin quantum number	I
nuclear spin quantum number [IUPAP]	I
number	n
number	N
number [British]	n
number, Avogadro's	N
number, Avogadro's [IUPAC]	L
number, Avogadro's [IUPAC]	N
number, cavitation [British]	σ
number density	n
number density of molecules [IUPAP]	n
number, dimensionless	N
number, Froude	F
number, Froude [British]	F
number, Froude [British]	Fr
number, Graetz [British]	Gz

number, Grashof	Gr
number, Grashof [British]	Gr
number in a sample [British]	n
number, Knudsen [British]	Kn
number, Mach	M
number, Mach [British]	M
number, Nusselt	Nu
number, Nusselt [British]	Nu
number of atoms, initial	N_0
number of atoms or nuclei at specified time	N
number of atoms per volume of a mixture [British]	N'
number of atoms per volume of a particular compound [British]	N
number of atoms per volume of a particular element [British]	N
number of components (Gibbs phase rule)	n
number of conductors	N
number of equivalents	J
number of lines of a grating, total	N
number of molecular collisions per unit time	Z
number of molecules	N
number of molecules [British]	N
number of molecules [IUPAC]	N
number of molecules [IUPAP]	N
number of molecules in a chemical reaction, stoichiometric	ν
number of molecules per unit volume	n
number of molecules per unit volume at zero degree centigrade and one atmosphere	n_0
number of molecules, stoichiometric [British]	ν
number of molecules, stoichiometric [IUPAC]	ν
number of molecules, total	N
number of moles	n
number of moles [British]	n
number of moles [IUPAC]	n
number of neutrons	N
number of neutrons [British]	n
number of neutrons in nucleus [British]	N
number of nuclei, initial	N_0
number of objects	n
number of observations	n
number of phases	m

number of phases [IEC]	m
number of revolutions per unit time	n
number of revolutions per unit time [IEC]	n
number of rotations per unit time	n
number of turns	N
number of turns [British]	N
number of turns [IUPAP]	N
number of turns in a winding [IEC]	N
number of turns per unit length	N
number, Peclet [British]	Pe
number, Prandtl	N_P
number, Prandtl	Pr
number, Prandtl	σ
number, Prandtl [British]	Pr
number, Rayleigh [British]	Ra
number, Reynolds	R
number, Reynolds	Re
number, Reynolds [British]	R
number, Reynolds [British]	Re
number, Richardson	Ri
number, Stanton	St
number, Stanton [British]	St
number, symmetry [IUPAC]	σ
Nusselt number	Nu
Nusselt number [British]	Nu
object distance	s
object height	y
object length	y
object nadir angle	η^*
observed value	X
operator, differential	d
optical air mass	m
optical air mass (subscript)	m
optical attenuation	D
optical density	D
optical density [British]	d
optical length	Δ
optical path difference	Δ
optical rotation angle [IUPAC]	α
optical rotation, angle of	θ
optical transmittance	τ
optical tube length	Δ

orbital angular momentum quantum number [IUPAP]	l_i
orbital angular momentum quantum number [IUPAP]	L
orbital magneton	μ_0
orbit inclination	ι
order of overtone	i
order of spectrum	m
order of spectrum	n
origin	O
orthogonal coordinates, Cartesian	x, y, z
orthogonal coordinates, Cartesian	X, Y, Z
oscillation period	T
oscillator figure of merit	M
osmotic coefficient [British]	g
osmotic coefficient [IUPAC]	g
osmotic coefficient [IUPAC]	φ
osmotic coefficient [IUPAP]	g
osmotic coefficient [IUPAP]	φ
osmotic pressure	p
osmotic pressure	Π
osmotic pressure [British]	Π
osmotic pressure [IUPAC]	Π
osmotic pressure [IUPAP]	Π
output (subscript)	o
output admittance	y_o
output conductance	g_o
output impedance	Z_o
output power	P_o
outside (subscript)	o
over-all heat transfer coefficient	U
over-all heat transfer coefficient [British]	U
over-all surface friction coefficient [British]	C_f
overpotential [British]	η
overpotential [IUPAC]	η
packing fraction [IUPAP]	f
partial capacitance coefficient	c
partial molal Gibbs function	μ
partial potential coefficient	p
partial pressure	p
particle displacement	ξ

DPMA

particle displacement in the x direction, component of	ξ
particle displacement in the y direction, component of	η
particle displacement in the z direction, component of	ζ
particle velocity, average	u_a
particle velocity, instantaneous	u_i
particle velocity in x direction, component of	u
particle velocity in y direction, component of	v
particle velocity in z direction, component of	w
particle velocity, maximum	u_m
particle velocity, peak	u_p
particle velocity, root-mean-square	u
partition function [British]	Q
partition function [IUPAC]	Q
partition function [IUPAP]	Q
partition function [IUPAP]	Z
path [IUPAC]	s
path [IUPAP]	L
path [IUPAP]	s
path, curved [IEC]	s
path difference, optical	Δ
path length	s
peak current	$\hat{I}$
peak current	I_p
peak current	I_{pk}
peak particle velocity	u_p
peak potential difference	$\hat{V}$
peak potential difference	V_p
peak potential difference	V_{pk}
peak potential difference, maximum	$\hat{V}_m$
peak power	P_p
peak sound pressure	p_p
Peclet number [British]	Pe
Peltier coefficient	Π
Peltier potential difference	V_π
perimeter	P
period	T
period	τ
period [British]	T
period [IUPAP]	T
periodicity	ω

periodicity, resonant	ω_r
periodic time [ISO]	T
periodic time [IUPAC]	T
periodic time [IUPAC]	τ
period of a periodic motion	T
period of decay [British]	τ
period of growth [British]	τ
permeability	μ
permeability [British]	μ
permeability [IEC]	μ
permeability [IUPAP]	μ
permeability, absolute	μ
permeabilty, magnetic	μ
permeability, magnetic [IUPAC]	μ
permeability of free space [British]	μ_0
permeability of free space [IEC]	μ_0
permeability of free space, reciprocal	μ_0
permeability of vacuum [IUPAP]	μ_0
permeability, reciprocal	ν
permeability. relative [British]	μ_r
permeability, relative [IUPAP]	μ_r
permeability, relative magnetic	μ
permeability, relative magnetic	μ_r
permeance	Λ
permeance [British]	Λ
permeance [IEC]	P
permeance [IEC]	Λ
permittance	c
permittance	C
permittivity	ϵ
permittivity [British]	ϵ
permittivity [IEC]	ε
permittivity [IEC]	ϵ
permittivity [IUPAC]	ϵ
permittivity [IUPAP]	ϵ
permittivity of free space	ϵ_0
permittivity of free space [British]	ϵ_0
permittivity of free space [IEC]	ε_0
permittivity of free space [IEC]	ϵ_0
permittivity of vacuum [IUPAP]	ϵ_0
permittivity, relative	ϵ_r
permittivity, relative [IUPAP]	ϵ_r

DPMA

perpendicular (subscript)	n
perturbation velocity, x component of	u'
perturbation velocity, y component of	v'
perturbation velocity, z component of	w'
phase angle	θ
phase angle	φ
phase angle	ϕ
phase angle	ψ
phase-change coefficient [British]	β
phase coefficient [IEC]	b
phase coefficient [IEC]	β
phase coefficient [ISO]	β
phase constant	β
phase difference [British]	ϕ
phase difference [IEC]	φ
phase difference [IEC]	ϕ
phase number [IUPAP]	m
phase velocity	u_φ
phase velocity	w
phase velocity of electromagnetic waves [British]	v
photoelectric threshold frequency	ν_0
photographic plate, inertia of	i
photometric brightness	B
photometric brightness [British]	L
photon [IUPAP]	γ
piezoelectric stiffness constant	h_{33}
piezoelectric strain constant	d
piezoelectric strain constant	δ
piezoelectric strain constant, hydrostatic	d_h
piezoelectric strain constant, planar	d_p
piezoelectric strain modulus	δ
piezoelectric stress constant	e
piezoelectric stress constant	ϵ
pion [IUPAP]	π
planar coupling factor (piezoelectricity)	k_p
planar piezoelectric strain constant	d_p
Planck's constant	h
Planck's constant [British]	h
Planck's constant [IEC]	h
Planck's constant [IUPAC]	h
Planck's constant [IUPAC]	h
Planck's constant [IUPAP]	h
Planck's function	Ψ

Planck's radiation law constant	c_1
Planck's radiation law constant	c_2
plane angle	θ
plane angle [IUPAP]	α
plane angle [IUPAP]	β
plane angle [IUPAP]	γ
plane angle [IUPAP]	θ
plane angle [IUPAP]	ϑ
plane angle [IUPAP]	φ
plate (subscript)	p
plate admittance	y_p
plate conductance	g_p
plate conductance at zero frequency and constant grid potential	k_p
plate power	P_p
plate resistance	r_p
Poisson's constant	κ
Poisson's ratio	μ
Poisson's ratio	ν
Poisson's ratio	σ
Poisson's ratio [British]	ν
Poisson's ratio [British]	σ
polarizability [IUPAP]	α
polarizability [IUPAP]	γ
polarizability of a molecule, electric [IUPAC]	α
polarizability of a molecule, electric [IUPAC]	γ
polarization, dielectric [IUPAC]	P
polarization, dielectric [IUPAP]	P
polarization, dielectric [IUPAP]	P
polarization, electric	P
polarization, electric [British]	P
polarization, magnetic	M
polarization, magnetic [British]	I
polarization, magnetic [British]	M
polarization, magnetic [IUPAP]	J
polarization, surface	P_s
polarizing angle of a dielectric material	$\bar{\varphi}$
polar moment of inertia	J
pole strength, magnetic	m
pole strength, magnetic	p
polytropic exponent	n

polytropic index [British]	n
porosity	P
position vector	$\mathbf{r}$
potential [IEC]	V
potential (subscript)	p
potential, chemical [British]	μ
potential, chemical [IUPAC]	μ
potential, chemical [IUPAP]	μ
potential coefficient, partial	p
potential difference	E
potential difference	V
potential difference [IEC]	U
potential difference [IEC]	V
potential difference, average	$\bar{V}$
potential difference, contact or Volta	V_v
potential difference, effective	V
potential difference, excitation	V_e
potential difference, instantaneous	v
potential difference, instantaneous [British]	v
potential difference, maximum	V_m
potential difference, maximum peak	$\hat{V}_m$
potential difference, maximum peak	V_{mp}
potential difference, peak	$\hat{V}$
potential difference, peak	V_p
potential difference, peak	V_{pk}
potential difference, Peltier	V_π
potential difference, quiescent	$\bar{V}$
potential difference, root-mean-square	V
potential difference, Seebeck	V_s
potential difference, steady direct-current	V
potential difference, Thomson	V_t
potential, electric	V
potential, electric [British]	V
potential, electric [IUPAC]	V
potential, electric [IUPAP]	V
potential, electric [IUPAP]	Φ
potential, electrokinetic [British]	ζ
potential, electrokinetic [IUPAC]	ζ
potential, electromagnetic scalar	φ
potential energy	E_p
potential energy	U
potential energy	V

potential energy [British]	V
potential energy [IUPAP]	E_p
potential energy [IUPAP]	V
potential head [British]	z
potential, inner	V
potential, inner	V_i
potential, ionization	V_i
potential, kinetic	L
potential, magnetic	$\mathcal{F}$
potential, magnetic scalar	$\mathcal{F}$
potential temperature	Θ
potential, velocity	φ
potential, velocity	ϕ
potential, velocity [British]	ϕ
power	P
power [British]	P
power [IEC]	P
power [IUPAC]	P
power [IUPAP]	P
power, acoustical	P_A
power, active	P
power, active [IEC]	P
power, apparent	P_a
power, apparent [British]	S
power, apparent [IEC]	P_s
power, apparent [IEC]	S
power, dispersive	ω
power dissipated per unit volume (piezoelectricity)	H
power, electrical	P_E
power factor (sinusoidal quantities) [British]	$\cos \phi$
power, input	P_i
power, instantaneous	p
power, mechanical	P_M
power of lens system	D
power, output	P_o
power, peak	P_p
power, plate	P_p
power, primary	P_p
power, radiant	P
power, radiant	Φ
power, radiant [IUPAC]	Φ
power, reactive	P_q

power, reactive [British]	Q
power, reactive [IEC]	P_q
power, reactive [IEC]	Q
power, refracting	D
power, rotational	P_R
power, secondary	P_s
power, thermoelectric	Q
Poynting vector	S
Poynting vector	Π
Poynting vector [British]	S
Poynting vector [IUPAP]	S
Poynting vector [IUPAP]	S
Prandtl number	N_P
Prandtl number	Pr
Prandtl number	σ
Prandtl number [British]	Pr
precipitable water	W
pressure	p
pressure [British]	p
pressure [IEC]	p
pressure [IUPAC]	p
pressure [IUPAC]	P
pressure [IUPAP]	p
pressure altitude	z_p
pressure, ambient	P_a
pressure, amplitude of simple harmonic	P
pressure, atmospheric	p
pressure, atmospheric	p_a
pressure, atmospheric [British]	p_a
pressure, atmospheric [British]	p_{at}
pressure, atmospheric [British]	P_{at}
pressure, average sound	p_a
pressure, coefficient of [British]	C_p
pressure, constant (subscript)	p
pressure, constant (subscript) [British]	p
pressure, critical	p_c
pressure head	h_p
pressure head [British]	h_p
pressure, instantaneous sound	p_i
pressure, maximum sound	p_m
pressure, osmotic	p
pressure, osmotic	Π

DPMA

pressure, osmotic [British]	Π
pressure, osmotic [IUPAC]	Π
pressure, osmotic [IUPAP]	Π
pressure, partial	p
pressure, peak sound	p_p
pressure ratio	δ
pressure ratio [British]	r_p
pressure, reduced	p_r
pressure, root-mean-square sound	p
pressure, specific	p
pressure, static	P_o
pressure, total	P
pressure, total [British]	p_o
pressure, total [British]	p_t
pressure, total [British]	P_o
pressure, total [British]	P_t
pressure, vapor	p
pressure, vapor	p^*
pressure, varying	p
pressure, water vapor	e
primary power	P_p
principal angle of incidence	$\bar{\varphi}$
principal angle of incidence of a dielectric material	$\bar{\varphi}$
principal azimuth angle	$\bar{\psi}$
principal quantum number [IUPAP]	n
principal quantum number [IUPAP]	n_i
probability	P
probability, resonance escape [British]	p
processes, isothermal (subscript)	θ
product [IUPAP]	Π
production rate, entropy	σ
product of inertia	I_{xy}
projected area	S
propagation coefficient [British]	γ
propagation coefficient [IEC]	p
propagation coefficient [IEC]	γ
propagation coefficient [ISO]	γ
propagation constant	γ
propagation flux density	Ξ
proton	p
proton [IUPAP]	p
proton mass [IUPAP]	m_p

proton number [IUPAP]	P
proton number [IUPAP]	Z
pseudo-adiabatic processes (with loss of condensate) (subscript)	s
pseudoscalar coupling [IUPAP]	P
pulsatance	ω
pulsatance [IUPAP]	ω
Q-factor [British]	Q
quadrupole moment [IUPAP]	Q
quadrupole moment, electric	Q
quality factor of reactor	Q
quantity	Q
quantity of electric charge [IEC]	Q
quantity of electricity [IEC]	Q
quantity of electricity [IUPAC]	Q
quantity of electricity [IUPAP]	Q
quantity of heat	Q
quantity of heat [British]	q
quantity of heat [British]	Q
quantity of heat [IUPAP]	Q
quantity of heat, aggregate [British]	Q
quantity of light	Q
quantity of light [IUPAC]	Q
quantity of light [IUPAP]	Q
quantity of radiant energy [IUPAP]	Q_e
quantum number, azimuthal or orbital	l
quantum number, hyperfine	F
quantum number, hyperfine [IUPAP]	F
quantum number, inner	j
quantum number, magnetic	m
quantum number, magnetic [IUPAP]	m_i
quantum number, magnetic [IUPAP]	M
quantum number, nuclear spin	I
quantum number, nuclear spin [IUPAP]	I
quantum number, orbital angular momentum [IUPAP]	l_i
quantum number, orbital angular momentum [IUPAP]	L
quantum number, principal	n
quantum number, principal [IUPAP]	n
quantum number, principal [IUPAP]	n_i
quantum number, rotational	R

DPMA

quantum number, rotational [IUPAP]	J
quantum number, rotational [IUPAP]	K
quantum number, spin	s
quantum number, spin [IUPAP]	s_i
quantum number, spin [IUPAP]	S
quantum number, total angular momentum [IUPAP]	j_i
quantum number, total angular momentum [IUPAP]	J
quantum number, total azimuthal or orbital	L
quantum number, total inner	J
quantum number, total magnetic	M
quantum number, total spin	S
quantum number, vibrational	v
quantum number, vibrational [IUPAP]	v
quiescent current	I
quiescent potential difference	$\bar{V}$
radial (subscript)	r
radial component of velocity	v
radial distance	r
radiance	N
radiance	$\mathcal{R}$
radiance [IUPAP]	L_e
radiance, spectral	N_λ
radiancy	$\mathcal{R}$
radiancy	W
radiant absorption	α
radiant emittance	W
radiant emittance [IUPAP]	M_e
radiant emittance, spectral	W_λ
radiant energy	U
radiant energy, spectral	U_λ
radiant flux	P
radiant flux	ϕ
radiant flux	Φ
radiant flux [IUPAP]	P
radiant flux [IUPAP]	Φ_e
radiant flux density	$\mathcal{E}$
radiant flux density	H
radiant flux density	$\mathcal{R}$
radiant flux density	W
radiant intensity	J

radiant intensity [IUPAP]	I_e
radiant intensity, extraterrestrial solar	I_0
radiant intensity, extraterrestrial solar	J_0
radiant intensity, spectral	J_λ
radiant intensity, total solar and sky	Q
radiant power	P
radiant power	Φ
radiant power [IUPAC]	Φ
radiant power, spectral	P_λ
radiant power which is absorbed, fraction of incident [IUPAC]	α
radiant power which is reflected, fraction of incident [IUPAC]	ρ
radiant power which is transmitted, fraction of incident [IUPAC]	τ
radiant reflectance	ρ
radiant transmittance	τ
radiation (subscript)	R
radiation (subscript) [British]	r
radiation frequency	ν
radiation, intensity of	I
radiation, intensity of [British]	I
radiation wavelength	λ
radiological decay constant	λ_r
radius	r
radius [British]	r
radius [IEC]	r
radius [ISO]	r
radius [IUPAC]	r
radius [IUPAP]	r
radius (subscript)	r
radius, Bohr	a_1
radius, Bohr [IUPAP]	a_0
radius of a diaphragm	a
radius of a disk	a
radius of a membrane	a
radius of a tube	a
radius of circle of least confusion	Z
radius of curvature	r
radius of curvature	ρ
radius of deformation, Rossby	λ
radius of Earth	a

radius of Earth .... R
radius of gyration .... k
radius of gyration .... r
radius of gyration [British] .... k
radius of gyration [British] .... r
radius of image of a point .... Z
radius of lens zone .... h
radius vector .... $\mathbf{r}$
rainfall duration .... D
range .... R
range [British] .... w
range, linear [IUPAP] .... R
range, linear [IUPAP] .... R_1
rate, discharge .... q
rate, heat transfer .... q
rate of entropy change per unit volume due to
    irreversibility .... σ
ratio .... R
ratio, cross-contraction .... σ
ratio, density .... σ
ratio, directivity .... $R_{\bullet}$
ratio, grid control .... μ
ratio, gyromagnetic [IUPAP] .... γ
ratio of circumference to diameter of circle
    [British] .... π
ratio of circumference to diameter of circle [IEC] .. π
ratio of plasma kinetic pressure to magnetic
    pressure .... β
ratio of reactance to resistance .... Q
ratio of specific heat capacity at constant pressure
    to specific heat capacity at constant volume
    [British] .... γ
ratio of specific heats .... γ
ratio of specific heats [British] .... k
ratio of specific heats [British] .... γ
ratio of specific heats [IUPAC] .... γ
ratio of specific heats [IUPAC] .... κ
ratio of specific heats [IUPAP] .... γ
ratio of specific heats [IUPAP] .... κ
ratio of speed to speed of light .... β
ratio of the circumference of a circle to its diameter . π
ratio, pressure [British] .... r_p

ratio, volume [British]	r_v
Rayleigh number [British]	Ra
reactance	X
reactance [British]	X
reactance [IEC]	X
reactance [IUPAC]	X
reactance [IUPAP]	X
reactance, acoustical	x_A
reactance, acoustical	X_A
reactance, capacitive	X_C
reactance, electrical	x_E
reactance, electrical	X_E
reactance, inductive	X_L
reactance, mechanical	x_M
reactance, mechanical	X_M
reactance, rotational	x_R
reactance, rotational	X_R
reaction, degree of	α
reaction, degree of [IUPAC]	α
reaction energy [IUPAP]	Q
reaction, equilibrium constant of [British]	K
reaction, extent of [IUPAP]	ξ
reaction, extent of chemical	ξ
reaction, extent of chemical [IUPAC]	ξ
reaction rate, specific	k
reaction velocity	u
reaction velocity constant	k
reaction, vertical	V
reactive power	P_q
reactive power [British]	Q
reactive power [IEC]	P_q
reactive power [IEC]	Q
real part of z [IUPAP]	z'
reciprocal capacitance	S
reciprocal impedance	Y
reciprocal of dispersive power	ν
reciprocal of viscosity	φ
reciprocal permeability	ν
reciprocal permeability of free space	μ_0
recombination coefficient [IUPAP]	A
recombination, coefficient of	α
rectangular moment of inertia	I

reduced mass	μ
reduced mass [IUPAP]	μ
reduced pressure	p_r
reduced properties (subscript)	r
reference area for drag and lift	A
reference conditions (subscript)	o
reference state (subscript) [British]	0
reflectance	ρ
reflectance [British]	ρ
reflection factor	ρ
reflection factor [British]	ρ
reflection factor [IUPAC]	ρ
reflection factor [IUPAP]	ρ
reflectivity	ρ
refracting power	D
refraction, angle of	r
refraction, angle of	φ'
refraction index [IUPAC]	n
refractive index	n
refractive index	μ
refractive index [British]	n
refractive index [British]	μ
refractive index [IUPAP]	n
refractive index, group	n_g
refractive index, group	μ_g
refractivity [IUPAC]	r
relative (subscript) [IUPAP]	r
relative activity [IUPAC]	a
relative activity of substance B [IUPAP]	a_B
relative activity of substance X [British]	a_x
relative activity of substance X [British]	$\{X\}$
relative atomic mass [IUPAP]	A_r
relative density [British]	d
relative density [IUPAC]	d
relative density [IUPAP]	d
relative dielectric coefficient	ϵ
relative dielectric coefficient	ϵ_r
relative humidity	f
relative humidity	H_R
relative humidity	r
relative humidity	U
relative humidity [British]	H_R

relative humidity [British]	ϕ
relative magnetic permeability	μ
relative magnetic permeability	μ_r
relative permeability	μ_r
relative permeability [IUPAP]	μ_r
relative permittivity	ϵ_r
relative permittivity [IUPAP]	ϵ_r
relative viscosity	η_r
relative vorticity, vertical component of	ζ
relativity ratio	β
relaxation time	τ
reluctance	$\mathcal{R}$
reluctance [British]	R
reluctance [British]	S
reluctance [IEC]	R
reluctance [IEC]	R_m
reluctance [IEC]	$\mathcal{R}$
reluctivity	ν
reluctivity [British]	ν
remanent or bias displacement	D_0
resilience, coefficient of	e
resistance	r
resistance	R
resistance [British]	R
resistance [IEC]	R
resistance [IUPAC]	R
resistance [IUPAP]	R
resistance, acoustical	r_A
resistance, cathode	R_k
resistance, coefficient of thermal	α
resistance, electrical	r_E
resistance, electrical	R_E
resistance, grid	R_g
resistance, load	R_L
resistance, mechanical	r_M
resistance, mechanical	R_M
resistance, plate	r_p
resistance, rotational	r_R
resistance, rotational	R_R
resistance, specific	ρ
resistance, specific [British]	r
resistance, specific [British]	ρ

resistance, specific acoustical	r_{A1}
resistance, temperature coefficient of	α
resistance, thermal	R
resistivity	ρ
resistivity [British]	r
resistivity [British]	ρ
resistivity [IEC]	ρ
resistivity [IUPAC]	ρ
resistivity [IUPAP]	ρ
resistivity, electrical	ρ
resistivity, thermal	k^{-1}
resolving power of telescope, angular	α
resonance escape probability [British]	p
resonant frequency	f_r
resonant frequency	f_R
resonant periodicity	ω_r
response, acoustical	R_A
response, electrical	R_E
response, mechanical	R_M
response, rotational	R_R
responsivity	R
restitution, coefficient of	e
rest length	l_0
rest mass	m_0
restoring force per unit displacement	k
resultant force	F_r
reverberation time	T
revolutions per unit time	n
Reynolds number	R
Reynolds number	Re
Reynolds number [British]	R
Reynolds number [British]	Re
Richardson number	Ri
right (subscript)	R
right-angle turning operator	i
right-angle turning operator	j
right-angle turning operator [IEC]	i
right-angle turning operator [IEC]	j
rigidity	n
rigidity, modulus of [British]	G
rolling friction, coefficient of	μ_r
root (subscript)	r

root-mean-square current	I
root of a Bessel equation	η_1
Rossby radius of deformation	λ
rotational (subscript)	r
rotational admittance	y_R
rotational admittance	Y_R
rotational compliance	C_R
rotational conductance	g_R
rotational conductance	G_R
rotational frequency	n
rotational frequency [British]	n
rotational frequency [IEC]	n
rotational frequency [ISO]	n
rotational impedance, complex	z_R
rotational impedance, complex	Z_R
rotational power	P_R
rotational quantum number [IUPAP]	J
rotational quantum number [IUPAP]	K
rotational reactance	x_R
rotational reactance	X_R
rotational resistance	r_R
rotational resistance	R_R
rotational response	R_R
rotational speed	n
rotational susceptance	b_R
rotational susceptance	B_R
rotational temperature	Θ_r
rotational temperature [IUPAP]	Θ_r
rotation, angle of optical	θ
rotation of Earth, angular speed of	Ω
rotation, specific	α
rotation, specific magnetic	V
rotation, specific magnetic	ω
roughness parameter	z_0
Rydberg constant	R
Rydberg constant [IUPAP]	R_∞
Rydberg constant for infinite mass	R_∞
Rydberg's constant [British]	R
sag of beam	δ
salinity	S
sample size	n
saturation	S

saturation (subscript)	s
scalar coupling [IUPAP]	S
scalar potential, electromagnetic	φ
scalar potential, magnetic	$\mathcal{F}$
scale of turbulence [British]	l
scattering angle [IUPAP]	θ
scattering angle [IUPAP]	ϑ
scattering angle [IUPAP]	φ
scattering coefficient	s
scattering cross-section	σ_s
scattering cross-section	Σ_s
secondary power	P_s
second moment of area [British]	I
second virial coefficient	B
second zonal harmonic coefficient	J
section modulus	Z
section modulus [British]	Z
sedimentation coefficient	S
Seebeck potential difference	V_s
self energy	ϵ
self-impedance	z
self-impedance	Z
self-inductance	L
self-inductance [British]	L
self-inductance [IEC]	L
self-inductance [IUPAC]	L
self-inductance [IUPAP]	L
semichord	b
semilatus rectum of conic section	p
semimajor axis	a
semiminor axis	b
sensible heat [British]	q_s
sensible heat [British]	Q_s
sensitivity of a phototube, dynamic	s
sensitivity of a phototube, static	S
shear coupling factor (piezoelectricity)	k_{15}
shear elasticity	μ
shear force [British]	Q
shearing force in beam section	V
shear modulus [British]	G
shear modulus [IUPAC]	G
shear modulus [IUPAP]	G

shear modulus of elasticity	G
shear modulus of elasticity	n
shear strain	γ
shear strain [British]	γ
shear strain [British]	ϕ
shear stress	s_s
shear stress	τ
shear stress [British]	q
shear stress [British]	τ
shear stress [IUPAC]	τ
shear stress [IUPAP]	τ
shear stress in fluid [British]	τ
shear velocity [British]	U_τ
shear velocity [British]	v^*
shell, strength of magnetic	I
shunt capacitance constant	Γ_0
single vortex, strength of	Γ
sliding friction, coefficient of	μ_k
sliding friction, coefficient of [British]	μ
slip (electrical machinery)	s
slip (electrical machinery)	σ
slip [IEC]	g
slip [IEC]	s
slit width	a
slope [British]	θ
solar flux, decimetric	F_{10}
solid (subscript)	X
solid (subscript) [British]	S
solid angle	ω
solid angle	Ω
solid angle [British]	ω
solid angle [British]	Ω
solid angle [IEC]	ω
solid angle [IEC]	Ω
solid angle [ISO]	ω
solid angle [ISO]	Ω
solid angle [IUPAC]	ω
solid angle [IUPAP]	ω
solid angle [IUPAP]	Ω
Sommerfeld constant	α
sound energy flux [IUPAP]	P
sound pressure, average	p_a

sound pressure, instantaneous .... p_i
sound pressure, maximum .... p_m
sound pressure, peak .... p_p
sound pressure, root-mean-square .... p
sound, speed of c
sound, speed of [British] .... a
sound, velocity of [IUPAP] c
sound, velocity of c
source (subscript) .... s
space, grating .... d
spacing of Bragg planes in a crystal .... d
specific (subscript) .... sp
specific absorbance [IUPAC] .... a
specific acoustical resistance .... r_{Al}
specific angular momentum .... h
specific conductance [British] .... γ
specific conductance [British] .... κ
specific conductance [British] .... σ
specific conductance [IUPAC] .... κ
specific enthalpy .... h
specific enthalpy [British] .... h
specific entropy .... s
specific entropy [British] .... s
specific gamma-ray constant .... Γ
specific gas constant .... R
specific gas constant [British] .... R
specific gas constant [British] .... $\mathcal{R}$
specific heat [IEC] c
specific heat at constant pressure [IUPAC] .... c_p
specific heat at constant volume [IUPAC] .... c_v
specific heat capacity c
specific heat capacity [British] c
specific heat capacity [IUPAP] .... c_p
specific heat capacity [IUPAP] .... c_v
specific heat capacity at constant pressure .... c_p
specific heat capacity at constant pressure [British] c_p
specific heat capacity at constant volume .... c_v
specific heat capacity at constant volume [British] c_v
specific heats, ratio of .... γ
specific heats, ratio of [British] .... γ
specific heats, ratio of [IUPAC] .... γ
specific heats, ratio of [IUPAC] .... κ

specific heats, ratio of [IUPAP]	γ
specific heats, ratio of [IUPAP]	κ
specific humidity	q
specific inductive capacity	ϵ
specific inductive capacity	ϵ_r
specific internal energy	u
specific internal energy [British]	u
specific internal energy of gas [British]	e
specific internal energy of gas [British]	u
specific magnetic rotation	V
specific magnetic rotation	ω
specific magnetic susceptibility	χ
specific magnetization	σ
specific reaction rate	k
specific resistance	ρ
specific resistance [British]	r
specific resistance [British]	ρ
specific volume	v
specific volume [British]	v
specific volume [British]	V_m
specific volume [IUPAC]	v
specific weight	γ
specific weight [IEC]	γ
spectral (subscript)	λ
spectral irradiance	H_λ
spectral radiance	N_λ
spectral radiant emittance	W_λ
spectral radiant energy	U_λ
spectral radiant intensity	J_λ
spectral radiant power	P_λ
speed	u
speed	v
speed, angular	ω
speed at specified time	u
speed at specified time	u_t
speed at specified time	v
speed at specified time	v_t
speed, average	$\bar{u}$
speed, average	u_{av}
speed, average	$\bar{v}$
speed, average	v_{av}
speed, initial	u_o

DPMA

speed, initial	v_o
speed, linear	u
speed, linear	v
speed, most probable	c_o
speed, most probable	$\hat{u}$
speed, most probable	$\hat{v}$
speed, most probable	α
speed of light [British]	c
speed of light in empty space [IUPAP]	c
speed of light in vacuum	c
speed of light in vacuum [ISO]	c
speed of propagation	v
speed of rotation [IEC]	n
speed of rotation of Earth, angular	Ω
speed of sound	c
speed of sound [British]	a
speed, particle	u
speed, particle	v
speed, root-mean-square	$\bar{u}$
speed, root-mean-square	$\bar{v}$
spherical coordinates	λ, ϕ, r
spin	s
spin, isotopic [British]	T
spin, nuclear	I
spin quantum number [IUPAP]	s_i
spin quantum number [IUPAP]	S
spin, total	S
spring constant	k
squareness ratio	R_s
square root of minus one	i
square root of minus one	j
standard acceleration due to gravity	g_o
standard acceleration due to gravity [British]	g_n
standard deviation	σ
standard deviation of a distributed variate [British]	σ
standard deviation of sample	s
standard gravitational acceleration [IUPAP]	g_n
standard or limiting condition (subscript) [British]	0
standard value (subscript)	o
standard value (subscript) [British]	n
Stanton number	St

Stanton number [British]	St
starting friction, coefficient of	μ_s
state, critical (subscript)	c
static density of the medium	ρ_0
static friction, coefficient of	μ_s
static sensitivity of a phototube	S
statistical weight	g
statistical weight [IUPAC]	g
statistical weight [IUPAC]	p
Stefan-Boltzmann constant	σ
Stefan-Boltzmann constant [British]	σ
steradiance	$\mathcal{B}$
steradiance	N
stiffness	s
stiffness constant, piezoelectric	h_{33}
stoichiometric activity coefficient	γ
stoichiometric activity coefficient [British]	γ
stoichiometric number of molecules [British]	ν
stoichiometric number of molecules [IUPAC]	ν
stoichiometric number of molecules in a chemical reaction	ν
stopping power, atomic [IUPAP]	S_a
stopping power, linear [IUPAP]	S
stopping power, linear [IUPAP]	S_l
strain	S
strain, constant (superscript)	S
strain constant, hydrostatic piezoelectric	d_h
strain constant, piezoelectric	d
strain constant, piezoelectric	δ
strain constant, planar piezoelectric	d_p
strain, direct [British]	e
strain, direct [British]	ϵ
strain energy [British]	U
strain modulus, piezoelectric	δ
strain, normal	ϵ
strain, shear	γ
strain, shear [British]	γ
strain, shear [British]	ϕ
stream function	ψ
stream function [British]	ψ
streamline (subscript)	s
strength, magnetic pole	m

strength, magnetic pole	p
strength of a simple source	A
strength of a vortex	C
strength of double layer	$\mathbf{P}_v$
strength of magnetic shell	I
strength of single vortex	Γ
strength of source	Q
strength of surface double layer	$\mathbf{P}_s$
stress	s
stress concentration factor	K
stress, constant (superscript)	T
stress constant, piezoelectric	e
stress constant, piezoelectric	ϵ
stress in fluid, shear [British]	τ
stress, normal	s
stress, normal	σ
stress, normal [British]	f
stress, normal [British]	σ
stress, normal [IUPAP]	σ
stress, shear	s_s
stress, shear	τ
stress, shear [British]	q
stress, shear [British]	τ
stress, shear [IUPAC]	τ
stress, shear [IUPAP]	τ
stretching force in membrane	F
stretching force in string	F
sublimation (subscript) [British]	s
sublimation process (subscript)	s
subsonic compressibility factor [British]	β
summation	Σ
summation [IEC]	Σ
summation [IUPAP]	Σ
supercompressibility factor	y
supersonic compressibility factor [British]	B
supersonic compressibility factor [British]	β
suppressor grid (subscript)	su
surface (subscript)	s
surface area	A
surface area	S
surface area per unit volume	a
surface area per unit volume	A_v

surface charge density	σ
surface charge density [British]	σ
surface charge density [IUPAP]	σ
surface concentration [British]	Γ
surface concentration [IUPAC]	Γ
surface density	ρ
surface density	σ
surface density of charge [IEC]	σ
surface emissivity	ϵ
surface friction coefficient, local [British]	c_f
surface friction coefficient, over-all [British]	C_f
surface heat transfer coefficient	h
surface in contact with substance (subscript) [British]	s
surface polarization	$\mathbf{P}$
surface tension	γ
surface tension	σ
surface tension [British]	γ
surface tension [British]	σ
surface tension [IUPAC]	γ
surface tension [IUPAC]	σ
surface tension [IUPAP]	γ
surface tension [IUPAP]	σ
surge impedance	Z_0
surge impedance	Z_0
susceptance	b
susceptance	B
susceptance [British]	B
susceptance [IEC]	B
susceptance [IUPAP]	B
susceptance, acoustical	b_A
susceptance, acoustical	B_A
susceptance, electrical	b_E
susceptance, electrical	B_E
susceptance, mechanical	b_M
susceptance, mechanical	B_M
susceptance, rotational	b_R
susceptance, rotational	B_R
susceptibility	κ
susceptibility	χ
susceptibility [IEC]	κ
susceptibility, electric	η

susceptibility, electric	χ_e
susceptibility, electric [IUPAP]	χ_e
susceptibilty, magnetic	χ_m
susceptibility, magnetic [IUPAC]	χ
susceptibility, magnetic [IUPAP]	χ_m
susceptibility, mass magnetic [British]	χ
susceptibility, specific magnetic	χ
susceptibility, volume magnetic	k
susceptibility, volume magnetic [British]	κ
Sutherland's constant	S
symmetry number [IUPAC]	σ
system rating constant	G_x
Tait free path	l_T
Tait free path	λ_T
tangential component (subscript)	t
temperature	t
temperature	T
temperature [IEC]	t
temperature [IEC]	θ
temperature [IEC]	ϑ
temperature [IUPAC]	t
temperature [IUPAC]	θ
temperature [IUPAP]	t
temperature [IUPAP]	ϑ
temperature, absolute	T
temperture, absolute	Θ
temperature, absolute [British]	T
temperature, absolute [IEC]	T
temperature, absolute [IEC]	Θ
temperature, absolute [IUPAC]	T
temperature, absolute virtual	T^*
temperature, absolute virtual	T_v
temperature, characteristic [IUPAC]	Θ
temperature, characteristic [IUPAP]	Θ
temperature coefficient [IEC]	α
temperature coefficient of resistance	α
temperature, constant (subscript)	T
temperature, constant (subscript) [British]	T
temperature, critical	t_c
temperature, critical	T_c
temperature, Curie	T_c
temperature, customary [British]	t

DPMA

temperature, customary [British]	θ
temperature, Debye [IUPAP]	Θ_D
temperature, Debye characteristic	θ_D
temperature, Debye characteristic	Θ
temperature, dew-point	τ
temperature difference	Δt
temperature, Einstein [IUPAP]	Θ_E
temperature, empirical [British]	t
temperature, empirical [British]	θ
temperature of ice point, ordinary	t_0
temperature, ordinary	t
temperature, ordinary	θ
temperature, potential	Θ
temperature, rotational	Θ_r
temperature, rotational [IUPAP]	Θ_r
temperature, thermodynamic	T
temperature, thermodynamic [IUPAP]	T
temperature, thermodynamic [IUPAP]	Θ
temperature, total (absolute) [British]	T_t
temperature, vibrational	Θ_v
temperature, vibrational [IUPAP]	Θ_v
tension in membrane	F
tension in string	F
tension, surface	γ
tension, surface	σ
tension, surface [British]	γ
tension, surface [British]	σ
tension, surface [IUPAC]	γ
tension, surface [IUPAC]	σ
thermal coefficient of volumetric expansion	β
thermal conductance	C
thermal conductivity	k
thermal conductivity	λ
thermal conductivity [British]	k
thermal conductivity [British]	λ
thermal conductivity [IUPAC]	λ
thermal conductivity [IUPAP]	λ
thermal diffusion coefficient	D_T
thermal diffusion coefficient [IUPAP]	D_T
thermal diffusion factor [IUPAP]	α_T
thermal diffusion ratio	k_T
thermal diffusion ratio [IUPAP]	K_T

thermal diffusivity	α
thermal diffusivity [British]	a
thermal diffusivity [British]	α
thermal diffusivity [British]	κ
thermal expansion, coefficient of	β
thermal expansion, cubic coefficient of [British]	γ
thermal expansion, linear coefficient of [British]	α
thermal expansion, total	δ
thermal neutron density [British]	n
thermal resistance	R
thermal resistivity	k^{-1}
thermodynamic temperature	T
thermodynamic temperature [IUPAP]	T
thermodynamic temperature [IUPAP]	Θ
thermoelectric power	Q
thickness	h
thickness	y
thickness [ISO]	d
thickness [ISO]	δ
thickness coupling factor (piezoelectricity)	k_t
Thomson coefficient	σ
Thomson potential difference	V_t
threshold frequency, photoelectric	ν_0
thrust	F
thrust (subscript)	T
thrust, horizontal	H
time	t
time	τ
time [British]	t
time [IEC]	t
time [ISO]	t
time [IUPAC]	t
time [IUPAP]	t
time constant	τ
time constant [IEC]	T
time constant [IEC]	τ
time constant of an exponentially varying quantity [ISO]	τ
time-dependent variable in Hamilton-Jacobi equation	V
time-dependent variable in Hamilton-Jacobi equation	W

time-dependent wave function	ψ
time-independent wave function	u
time interval [IUPAC]	T
time interval [IUPAC]	τ
time of one cycle [IEC]	T
time, periodic [ISO]	T
time, periodic [IUPAC]	τ
time, relaxation	τ
time, reverberation	T
torque	T
torque [British]	T
torque [IEC]	T
torque per unit twist	k
torsion constant	k
total (subscript)	t
total acoustical absorption in a room	a
total angular momentum	H
total angular momentum quantum number [IUPAP]	j_i
total angular momentum quantum number [IUPAP]	J
total current	I_T
total dielectric flux	Ψ
total differential of x [IUPAP]	dx
total electric flux	Ψ
total elongation	δ
total emissivity [British]	ϵ
total enthalpy	H
total entropy	S
total entropy [British]	S
total heat [British]	H
total heat capacity	C
total impedance	Z_T
total inner quantum number	J
total internal energy	U
total load	P
total load	W
total moles	n
total pressure	P
total pressure [British]	p_0
total pressure [British]	p_t
total pressure [British]	P_0
total pressure [British]	P_t
total solar and sky radiant intensity	Q

total (absolute) temperature [British]	T_t
total volume	V
total volume	τ
total vorticity	Z
traction [IUPAC]	σ
trajectory (subscript)	t
transadmittance, grid-plate	y_{pg}
transconductance	G_M
transconductance at zero frequency and constant plate potential	s_p
transconductance, grid-plate	g_m
transconductance, grid-plate	g_{pg}
transconductance, inverse	g_{gp}
transconductance, inverse	g_n
transition (subscript) [British]	t
transition between polymorphic forms (subscript)	t
transmission factor	τ
transmission factor [British]	τ
transmission factor [IUPAC]	τ
transmission factor [IUPAP]	τ
transmissivity	τ
transmittance	τ
transmittance [British]	τ
transmittance [IUPAC]	T
transmittance, optical	τ
transmittance, radiant	τ
transport number [British]	t
transport number [British]	T
transport number [IUPAC]	t
transport number [IUPAC]	T
transpose of matrix A [IUPAP]	$\tilde{A}$
transverse component of velocity	u
transverse coupling factor (piezoelectricity)	k_{31}
transverse electric field	E_h
triton	t
triton [IUPAP]	t
turbidity	s
turbulence-correlation coefficient	R
turbulence exchange coefficient	ϵ
turbulence scale [British]	l
turbulence wave number [British]	κ
turning operator, 90-degree	i
turning operator, 90-degree	j

unified atomic mass constant [IUPAP]	m_u
unit vector along X-axis	$\mathbf{i}$
unit vector along Y-axis	$\mathbf{j}$
unit vector along Z-axis	$\mathbf{k}$
unit vector tangent to path	$\boldsymbol{\tau}$
vacuum, permeability of [IUPAP]	μ_0
vacuum, permittivity of [IUPAP]	ϵ_0
valence	z
valency of an ion [British]	z
value, critical (subscript)	c
value of a quantity x, average	$\bar{x}$
Van't Hoff coefficient	i
vapor (subscript)	G
vapor density	ρ
vaporization processes (subscript)	v
vapor pressure	p
vapor pressure	p^*
vapor pressure [British]	p
vapor pressure constant	i
variance	σ^2
variation of x [IUPAP]	δx
vector along X-axis, unit	$\mathbf{i}$
vector along Y-axis, unit	$\mathbf{j}$
vector along Z-axis, unit	$\mathbf{k}$
vector coupling [IUPAP]	$\mathbf{V}$
vector in Z-direction, unit	$\mathbf{k}$
vector potential [IUPAP]	$\mathbf{A}$
vector potential, magnetic	$\mathbf{A}$
vector, Poynting	$\boldsymbol{\Pi}$
vector tangent to path, unit	$\boldsymbol{\tau}$
velocity	u
velocity	$\mathbf{u}$
velocity	v
velocity	$\mathbf{v}$
velocity [British]	v
velocity [ISO]	c
velocity [ISO]	u
velocity [ISO]	v
velocity [ISO]	w
velocity [IUPAC]	u
velocity [IUPAC]	v
velocity [IUPAC]	w
velocity [IUPAP]	c

velocity [IUPAP]	u
velocity [IUPAP]	v
velocity, angular	ω
velocity, angular	$\boldsymbol{\omega}$
velocity, angular	Ω
velocity, angular [British]	ω
velocity, angular [British]	Ω
velocity, angular [IEC]	ω
velocity, angular [ISO]	ω
velocity, angular [IUPAC]	ω
velocity, angular [IUPAP]	$\boldsymbol{\omega}$
velocity at time t	$\mathbf{u}$
velocity at time t	$\mathbf{u}_t$
velocity at time t	$\mathbf{v}$
velocity at time t	$\mathbf{v}_t$
velocity, average	$\bar{\mathbf{u}}$
velocity, average	$\mathbf{u}_{av}$
velocity, average	$\bar{\mathbf{v}}$
velocity, average	$\mathbf{v}_{av}$
velocity, average [IUPAP]	c_0
velocity, average [IUPAP]	u_0
velocity, average particle	u_a
velocity constant of reaction [British]	k
velocity constant, reaction	k
velocity distribution function [IUPAP]	$f(c)$
velocity, friction	u^*
velocity, friction [British]	U_τ
velocity, friction [British]	v^*
velocity, group	$\mathbf{u}$
velocity, group	$\mathbf{u}_g$
velocity, group	$\mathbf{v}$
velocity, group	$\mathbf{v}_g$
velocity head	h_v
velocity head [British]	h_v
velocity, initial	$\mathbf{u}_o$
velocity, initial	$\mathbf{v}_o$
velocity, instantaneous particle	u_i
velocity, ionic	u
velocity, linear	$\mathbf{u}$
velocity, linear	v
velocity, linear [British]	u
velocity, linear [British]	v

velocity, linear [British]	w
velocity, linear [IEC]	v
velocity, mass	G
velocity, maximum particle	u_m
velocity, most probable	c_0
velocity, most probable	$\hat{u}$
velocity, most probable	$\hat{v}$
velocity, most probable	α
velocity of light in vacuum	c
velocity of light in vacuum [IEC]	c
velocity of longitudinal waves [IUPAP]	c_l
velocity of sound [IUPAP]	c
velocity of sound or other waves	v
velocity of transverse waves [IUPAP]	c_t
velocity, particle	$\mathbf{u}$
velocity, particle	$\mathbf{v}$
velocity, peak particle	u_p
velocity, phase	u_φ
velocity, phase	w
velocity potential	φ
velocity potential	ϕ
velocity potential [British]	ϕ
velocity potential amplitude	A
velocity potential, complex [British]	w
velocity, reaction	u
velocity, root-mean-square	$\bar{u}$
velocity, root-mean-square	$\bar{v}$
velocity, shear [British]	U_τ
velocity, shear [British]	v^*
velocity, transverse component of	u
velocity, volume	U
velocity, wave	c
velocity, wave	u_w
velocity, wave	w
velocity, x component of	u
velocity, y component of	v
velocity, z component of	w
Verdet constant	V
Verdet constant	ω
vertical (subscript)	v
vertical axis	Y
vertical component of absolute vorticity	ζ_a

vertical component of relative vorticity	ζ
vertical force	F_v
vertical reaction	V
vibrational quantum number [IUPAP]	v
vibrational temperature	Θ_v
vibrational temperature [IUPAP]	Θ_v
virial coefficient, fourth	D
virial coefficient, second	B
virtual quantities (subscript)	v
viscosity [IUPAC]	η
viscosity [IUPAP]	η
viscosity, absolute	μ
viscosity, coefficient of	η
viscosity, coefficient of kinematic	ν
viscosity, coefficient of molecular	μ
viscosity, dissipative	η
viscosity, dynamic	η
viscosity, dynamic	μ
viscosity, dynamic [British]	η
viscosity, dynamic [British]	μ
viscosity, frictional	η
viscosity, kinematic	ν
viscosity, kinematic [British]	ν
viscosity, kinematic [IUPAC]	ν
viscosity, kinematic [IUPAP]	ν
viscosity, reciprocal of	φ
viscosity, relative	η_r
viscosity, volume	κ
visibility factor	K_λ
voltage	E
voltage	V
voltage, alternating-current filament supply	E_f
voltage, cutoff	E_{co}
voltage efficiency	η_v
volume	V
volume [British]	v
volume [British]	V
volume [IEC]	V
volume [ISO]	v
volume [ISO]	V
volume [IUPAC]	v
volume [IUPAC]	V
volume [IUPAP]	v

volume [IUPAP]	V
volume charge density [British]	ρ
volume, constant (subscript)	v
volume, constant (subscript) [British]	v
volume, critical	V_c
volume current	U
volume density of charge [IEC]	ρ
volume density of electric charge	ρ
volume displacement	X
volume expansivity	β
volume flow rate [British]	Q
volume fraction of substance B [IUPAP]	φ_B
volume in gamma phase space [IUPAP]	Ω
volume magnetic susceptibility [British]	κ
volume modulus of elasticity	B
volume of a cavity or room	V
volume of phase space	Ω
volume per atom	v
volume per atom or molecule	v_m
volume per mole	v
volume per mole	V
volume per mole	V_M
volume per molecule	v
volume per unit mass	v
volume ratio [British]	r_v
volume, specific	v
volume, specific [British]	v
volume, specific [British]	V_m
volume, specific [IUPAC]	v
volume susceptibility, magnetic	k
volume, total	V
volume, total	τ
volumetric diffusivity	D_v
volumetric expansion, thermal coefficient of	β
volumetric flow rate	Q
volume velocity	U
volume viscosity	κ
Von Karman's constant	k_0
vortex strength [British]	K
vortex strength [British]	Γ
vortex, strength of a	C
vortex, strength of single	Γ
vorticity [British]	ω

vorticity, east-west component of horizontal relative	ξ
vorticity, north-south component of horizontal relative	η
vorticity, total	Z
vorticity, vertical component of absolute	ζ_0
vorticity, vertical component of relative	ζ
wall (subscript)	w
wall (subscript) [British]	w
water, precipitable	W
water vapor (subscript)	w
water-vapor content	w
water vapor pressure	e
wave function, time-dependent	ψ
wave function, time-independent	u
wavelength	λ
wavelength [British]	λ
wavelength [ISO]	λ
wavelength [IUPAC]	λ
wavelength [IUPAP]	λ
wavelength (subscript)	λ
wavelength, Compton [IUPAP]	λ_C
wavelength constant	k
wavelength, effective	λ_e
wavelength, radiation	λ
wave number	k
wave number	σ
wave number [British]	ν
wave number [British]	$\tilde{\nu}$
wave number [ISO]	$\tilde{\nu}$
wave number [ISO]	σ
wave number [IUPAC]	ν
wave number [IUPAC]	σ
wave number [IUPAP]	$\tilde{\nu}$
wave number [IUPAP]	σ
wave number, circular	n
wave number, circular [ISO]	k
wave number, circular [IUPAP]	k
wave number, turbulence [British]	κ
wave velocity	c
wave velocity	u_w
wave velocity	w

weight	F
weight	w
weight	W
weight [British]	W
weight [IUPAC]	G
weight [IUPAC]	W
weight [IUPAP]	G
weight [IUPAP]	W
weight density [British]	w
weight, equivalent	M/z
weightivity	γ
weight, molecular	M
weight, molecular [British]	M
weight per unit volume	γ
weight, specific	γ
weight, statistical	g
weight, statistical [IUPAC]	g
weight, statistical [IUPAC]	p
Weiss constant	θ
wet-bulb temperature (subscript)	w
width	b
width	w
width, level [British]	Γ
width, level [IUPAP]	Γ
width of slit	a
Wien's displacement constant	b
work	w
work	W
work [British]	w
work [IEC]	A
work [IEC]	W
work [IUPAC]	A
work [IUPAC]	w
work [IUPAP]	A
work [IUPAP]	W
work function, gross	w_g
work function, net	w
work function per unit charge, gross	φ_g
work function per unit charge, net	φ
x component of perturbation velocity	u'
x component of velocity	u

x, y, z coordinates (subscript)	x, y, z
y component of perturbation velocity	v'
y component of velocity	v
Young's modulus [British]	E
Young's modulus [IUPAP]	E
Young's modulus of elasticity	E
Young's modulus of elasticity	Y
z component of perturbation velocity	w'
z component of velocity	w

MATHEMATICAL SIGNS AND SYMBOLS

GENERAL & ARITHMETIC

$+$	plus, addition, positive direction	
$-$	minus, subtraction, negative direction	
$=$	equals	
$\neq$	does not equal	
$\pm$	plus or minus	
$\mp$	minus or plus	
$\times$; $a \cdot b$; ab	multiplication	
$\div$; $\dfrac{a}{b}$; a/b	division	
$:$	ratio	
$>$	greater than	
$<$	less than	
$\geq$	greater than or equal to	
$\leq$	less than or equal to	
$\gtrless$	greater than or less than	
$\lesseqgtr$	equal to or less than	
$\gtreqless$	equal to or greater than	
$\gtreqqless$	greater than or equal to or less than	
$::$	proportionally equal to	
$\approx$	approximately equal to	
$\equiv$	identical to	
$\not\equiv$	not identical to	
$\not<$	not less than	
$\propto$	varies directly as	
∞	infinity	
$\rightarrow$	approaches as a limit	
$\lvert x \rvert$	absolute value of x	
$\log_a x$	logarithm of x to the base a	
$\log_e x$; $\ln x$	natural logarithm of x	
$\log_{10} x$; $\log x$	common logarithm of x	
G.C.D.; gcd	greatest common denominator	
(a, b)	greatest common divisor of a and b	
L.C.M.; lcm	least common multiple	
$[a, b]$	least common multiple of a and b	

DPMA

Symbol	Meaning
m/n	raised to the power of degree m/n (superscript)
$-n$	negative exponent (superscript)
i	imaginary unit
$\{\ \}$	notation for set
I	the universal set
aRb	a is in the relation R to b
$0.1428\dot{7}$	repeating decimal
$0.21\dot{3}84\dot{2}$	repeating sequence of decimals
(a, b)	open interval
$[a, b]$	closed interval
$(a, b]$; $[a, b)$	half-closed or half-open interval
$f(x)$	function of x
$[x]$	the greatest integer not exceeding x
$\{x\}$	the difference between x and the greatest integer not exceeding x
$\exists$	there exists
$\ni$	such that
$\forall$	for all
$\in$	belongs to
$\notin$	does not belong to
$\rightarrow$; $\Rightarrow$	implies
$\leftrightarrow$; $\Leftrightarrow$; iff	if and only if
$<\ >$	average
$-$	overbar, indicates average
Re	real part of
Im	imaginary part of
O	bound
$n!$; $\lfloor n$; $\Pi(n)$	factorial of n
$\div$	geometrical proportion
$\therefore$	therefore, hence
$\because$	because
Q.E.D.	which was to be demonstrated, or proved (*quod erat demonstrandum*)
$\sim$	similar to, proportional to
$++$	double plus
$+\cdots$	plus other terms
$+-\cdots$	plus other terms with alternating sign
$++--\cdots$	plus other terms, two positive alternating with two negative
$\underline{\underline{m}}$	is measured by
$\doteq$	equals in the limit

DPMA

$\sqrt{}$	radical, square root	
$\sqrt[3]{}$	cube root	
$\sqrt[n]{}$	nth root	
$-$	vinculum, as subscript, for grouping	
$O(x)$	of the order of x	
$o(x)$	of the small order of x	

ALGEBRA

$nPr; P_r^n$	permutations of n things taken r at a time
$nCr; C_r^n; \binom{n}{r}$	combination of n things taken r at a time
a_{ij}	element in row i, column j of determinant or matrix
$\det(a_{ij})$	determinant with elements a_{ij} or determinant of matrix (a_{ij})
i, j, k	vectors of unit magnitude
$\mathbf{A} \cdot \mathbf{B}$	scalar product of two vectors
$\mathbf{A} \wedge \mathbf{B}; \mathbf{A} \times \mathbf{B}$	vector product of two vectors
$\mathbf{AB}$	dyadic product of two vectors
$\mathbf{S} : \mathbf{T}$	scalar product of two tensors
$\mathbf{S} \cdot \mathbf{T}$	tensor product of two tensors
$\mathbf{S} \cdot \mathbf{A}$	product of tensor $\mathbf{S}$ and vector $\mathbf{A}$
$A \angle \theta$	indicates the vector $\mathbf{A} = a\mathbf{i} + b\mathbf{j}$ (or) $\mathbf{i}a + \mathbf{j}b$ where $a = \|\mathbf{A}\| \cos \theta$, $b = \|\mathbf{A}\| \sin \theta$, $\theta = \arctan b/a$, and $\mathbf{A} = (a^2 + b^2)^{\frac{1}{2}}$
∇	del, differential operator $= \mathbf{i}\dfrac{\partial}{\partial x} + \mathbf{j}\dfrac{\partial}{\partial y} + \mathbf{k}\dfrac{\partial}{\partial z}$
∇^2	Laplacian $= \dfrac{\partial^2}{\partial x^2} + \dfrac{\partial^2}{\partial y^2} + \dfrac{\partial^2}{\partial z^2}$
$r \operatorname{cis} \theta$	polar form of a complex number $= r(\cos \theta + i \sin \theta)$
$re^{i\theta}$	exponential form of complex number
a^{-1}	inverse element of a in a group
I	identity element in a group
$\oplus; \otimes$	operations in a field
$p(x)$	polynomial in x
$\tilde{A}$	transpose of matrix A
$A^\dagger$	conjugate matrix, adjoint matrix
A^{-1}	inverse matrix
$\|a\|$	absolute value of a
$\|\ \ \|$	rectangular matrix
$[\mathbf{abc}]$	triple product of three vectors

DPMA

λ	eigenvalues or characteristic values	
$\cdot$	dot or scalar product	
$\times$	cross product	
$\| \ \|$	determinant	
J	Jacobian	
$\dfrac{\partial(u_1, u_2, \cdots, u_n)}{\partial(x_1, x_2, \cdots, x_n)}$	Jacobian	
a^2	product $a \cdot a$	
a^n	product $a \cdot a \cdots a$ of n factors a	
$\sqrt{a}; a^{\frac{1}{2}}$	square root of a	
$\sqrt[n]{a}; a^{1/n}$	nth root of a	
a^0	unity, 1	
a^{-1}	reciprocal, $1/a$	
a^{-n}	reciprocal, $1/a^n$	
$a^{m/n}$	nth root of a^m	
q	quaternion	

GEOMETRIES

$\angle, \angle s$	angle(s)	
$\perp$	perpendicular to	
$\perp, \perp s$	perpendicular(s)	
$\parallel$	parallel to	
$\parallel, \parallel s$	parallel(s)	
$\triangle, \triangle s$	triangle(s)	
$\bigcirc, \bigcirc s$	circle(s)	
$\Box, \Box s$	parallelogram(s)	
$\Box, \Box s$	square(s)	
$\Box, \Box s$	trapezoid(s)	
$\cong; \equiv$	(is) congruent (to)	
$\sim$	(is) similar (to)	
$\stackrel{\vee}{=}$	(is) equiangular	
$\therefore$	therefore	
$\overline{AB}$	vinculum, chord AB of a circle or length of line segment between A and B	
$\overleftarrow{AB}$	directed segment B to A	
$\stackrel{\frown}{AB}$	arc AB of a circle	
π	pi, constant ratio of circumference to diameter of a circle	

DPMA

a.s.a.		angle, side, angle
s.a.s.		side, angle, side
s.s.s.		side, side, side
AB		line segment between A and B
$\vec{AB}$		vector, or directed line segment from A to B
$A°$		A degrees of angle
$\stackrel{m}{=}$		is measured by
E^n		Euclidean n-space
O		origin of coordinate system
X		axis of abscissas
Y		axis of ordinates
Z		third axis of coordinates in three-dimensional space

ANALYTIC GEOMETRY

x, y		rectangular coordinates of a point in a plane
x, y, z		rectangular (Cartesian) coordinates of a point in space
α		alpha, indicates direction angle with X-axis
l		indicates directional cosine (with X-axis)
β		beta, indicates direction angle with Y-axis
m		indicates directional cosine (with Y-axis)
γ		gamma, indicates direction angle with Z-axis
n		indicates directional cosine (with Z-axis)
r, θ		polar coordinates of a point in a plane
r, θ, ϕ		spherical coordinates of a point in space
ψ		psi, indicates angle from radius vector to tangent of plane curve
r, θ, z		cylindrical coordinates of a point in space
p, s		indicates intrinsic coordinates
e		eccentricity of a conic
p		semi-latus rectum
m		slope of a curve or line
C		circumference of a circle
r		radius of a circle
D		diameter of a circle
ρ		radius of curvature
d		perpendicular distance from a point to a line (length of normal)

TRIGONOMETRY

°	indicates degree(s)
θ	angle measured in radians
′	prime, indicates minutes
″	double prime, indicates seconds
sin	sine function
cos	cosine function
tan	tangent function
cot; ctn	cotangent function
sec	secant function
csc	cosecant function
vers	versed sine function = $(1-\cos\theta)$
covers	coversed sine function = $(1-\sin\theta)$
hav	haversine function = $\frac{1}{2}(1-\cos\theta)$
cis θ	$\cos\theta + i\sin\theta$
$\sin^{-1}$; arc sin	inverse sine function
$\cos^{-1}$; arc cos	inverse cosine function
$\tan^{-1}$; arc tan	inverse tangent function
$\sec^{-1}$; arc sec	inverse secant function
$\csc^{-1}$; arc csc	inverse cosecant function
$\cot^{-1}$; arc cot; ctn^{-1}; arc ctn	inverse cotangent function
sinh	hyperbolic sine function
cosh	hyberbolic cosine function, etc.
$\sinh^{-1}$; arc sinh	inverse hyperbolic sine function
$\cosh^{-1}$; arc cosh	inverse hyperbolic cosine function, etc.
l, m, n	direction numbers in space
$\cos\alpha, \cos\beta, \cos\gamma$	direction cosines in space

CALCULUS

Δ	delta, indicates increment
$\sum_i$	sigma, indicates summation, sum of terms of index i
$\sum_{i=1}^{n} a_i$	sum $a_1 + a_2 + \cdots + a_n$
$\{a_n\}; (a_n)$	infinite sequence $a_1, a_2, a_3 \cdots$
$\sum_{i=1}^{\infty} a_i$	sum of infinite series $a_1 + a_2 + a_3 \cdots$

$\lim\limits_{n \to \infty} a_n$ .... limit of a_n as n approaches infinity ....

$\lim\limits_{x=a} f(x); \lim\limits_{x \to a} f(x)$ .. limit of $f(x)$ as x approaches a ....

Δx .... increment in x ....

dx .... differential of x ....

$''$ .... double prime, second order of differentiation ......

$'''$ .... triple prime, third order of differentiation ....

$df(x)$ .... differential of $f(x) = f'(x)dx$ ....

$\dfrac{df(x)}{dx}; D_x f(x); f'(x)$. derivative of $f(x) = \lim\limits_{x \to 0} \dfrac{f(x+\Delta x) - f(x)}{\Delta x}$ ....

$\int, \iint, \iiint$ .... integral signs ....

$\int_i^m, \int_a^b, \int_a^b \int_c^d$ integral signs, indicating index and limits ....

$\oint$ .... integral around a closed path ....

$\int f(x)dx$ .... indefinite integral of $f(x)$ ....

$\int_a^b f(x)dx$ .... definite integral of $f(x)$ from a to b ....

$Fx|_a^b; Fx]_a^b$ ...... difference $F(b) - F(a)$ ....

$\dfrac{d^2 f(x)}{dx^2};$

$D_x^2 f(x); f''(x)$ ... second derivative of $f(x)$ ....

$\dfrac{d^n f(x)}{dx^n};$

$D_x^n f(x); f^n(x)$ .... nth derivative of $f(x)$ ....

$\dfrac{\partial f(x, y)}{\partial x};$

$\quad D_x f(x,y); f_x(x,y)$. partial derivative of $f(x,y)$ with respect to x ....

$f^{-1}(x)$ .... function inverse to the function $f(x)$ ....

$\dfrac{\partial^2 f(x, y)}{\partial y \partial x};$

$\quad D_{yx}^2 f(x,y);$

$\quad f_{yx}(x,y)$ .... second partial derivative of $f(x, y)$ with respect to x and then y ....

$d^n; D^n$ .... differential operator of nth order ....

$\dfrac{d}{dx}$ .... derivative operator of first order ....

$\dfrac{d^n}{dx^n}$ .... derivative operator of nth order ....

∂_x .... partial differential operator with respect to x ......

$\dot{x}$ .... first derivative with respect to time ....

$\ddot{x}$ .... second derivative with respect to time ....

$\dfrac{d^n y}{dx^n}$ .... derivative of nth order ....

$J_0(x), J_1(x), \cdots,$
$\quad J_n(x)$ .... Bessel functions (the notation recommended is G. N. Watson's treatise, 1922, as endorsed by E. P. Edams in the Smithsonian tables, 1922) ....

$B_1, B_2, B_3, \cdots$ .... Bernoulli numbers ....

γ .... gamma, Euler's (Mascheroni's) constant ($0.5772\cdots$).

$\Gamma(n) = \displaystyle\int_0^\infty x^{n-1} e^{-x} dx = \int_0^1 [\log(1/x)]^{n-1} dx$
.... Gamma function of n ....

$B(m, n) = \displaystyle\int_0^1 x^{m-1}(1-x)^{n-1} dx$
.... Beta function of any two positive numbers m and n

$\Gamma_x(n) = \displaystyle\int_0^n x^n e^{-n} dx$
.... incomplete Gamma function ....

$B_x(m, n) = \displaystyle\int_0^n x^{m-1}(1-x)^{n-1} dx$
.... incomplete Beta function ....

ω .... transfinite cardinal ....

$\wp$ .... Weierstrassian function ....

LOGIC, SET THEORY, AND TOPOLOGY

$\sim P; \overline{P}$ .... not P ....

$P \wedge Q; P \cdot Q;$
$\quad P \& Q$ .... P and Q ....

$P \vee Q; P + Q$ ..... P or Q ....

$P \to Q; P \supset Q$ ..... if P then Q ....

$a \in A$ .... a is a member of set A ....

$P \leftrightarrow Q; P \equiv Q;$
$\quad P \text{ iff } Q$ .... P if and only if Q ....

aRb .... a is in the relation R to b ....

$A = B$ .... set A is identical with set B ....

$A \subset B$ .... set A is included in set B ....

$A \supset B$ .... set A contains set B ....

$A \cap B; A \cdot B$ ..... intersection of sets A and B ....

$A \cup B; A + B$ .... union of sets A and B ....

DPMA

$CA; A'; \bar{A}$	complement of set A	
$\Lambda; 0; \phi$	null set	
$(x); (\forall x); A_x$	for every x	
$(\exists x); (Ex); Ex$	there exists an x such that	
$\aleph_0$	aleph null, cardinal number of positive integers	
c	cardinal number of real numbers	
ω	ordinal number of the positive integers	
I	the universal set	
R	set of real numbers	
E^n	Euclidean n-space	
c	complement of a set (superscript)	
$\|\ \|$	rectangular matrix	
λ	eigenvalues or characteristic values	
$\oint$	line integral along a closed path	
$\cdot$	dot or scalar product	
$\times$	cross product	
$\|\ \|$	determinant	
$\mathcal{L}$	Laplace transform	
J	Jacobian	
μ	set measure	
$\exists$	there exists	
$\exists !$	there exists a unique	
$\ni$	such that	
$\forall$	for all	
$\in$	belongs to	
$\notin$	does not belong to	
$\rightarrow; \Rightarrow$	implies	
$\leftrightarrow; \Leftrightarrow; iff$	if and only if	
$<\ >, av$	average	
$(a\|p)$	Legendre symbol	
I	identity matrix	
ϕ	null vector	
ϕ	null space	
$\{\ \}$	notation for set	
(a, b)	ordered pair	
(a, b)	open interval	
$[a, b]$	closed interval	
$(a, b]; [a, b)$	half-closed or half-open interval	
$b^{m/n}$	nth root of b^m	
i	imaginary unit	
$\cdot$	operation in a group	

a^{-1}	inverse element of a in a group	
I	identity element in a group	
$\oplus$; $\otimes$	operations in a field	
f	function	
$P(x)$; $P[x]$	polynomial in x	
[ ]	the greatest integer not exceeding the quantity inside the square brackets	
Re	real part of	
Im	imaginary part of	
O	bound	
$\cong$; $\simeq$	approximately equal to	
$+0$	reaching 0 from the positive or right side	
-0	reaching 0 from the negative or left side	
$f(t) \supset \phi(p)$	$\phi(p)$ is the ordinary Laplace transform of $f(t)$	
$f(t) = O(t^v)$	$\|f(t)\| < kt^v$, k being an absolute positive constant independent of variables or parameters	
$'$	first derivative	
$''$	second derivative	
n	nth derivative (superscript)	

STATISTICS & PROBABILITY

$P(x)$	probability distribution of x	
$p(x)$	probability density function of x	
$P(E)$	probability of the event E	
$P(x > A)$	probability that observation x is greater than the value A	
$P(A < x < B)$	probability that observation x will fall between the values A and B	
$\cap$	intersection	
$\cup$	union	
$\subset$	contained in	
$\supset$	contains	
$P(E \cap F)$	probability of E and F simultaneously occurring	
$P(E \cup F)$	probability of at least one of the events E and F occurring	
$P(E\|F)$; $P(E \cap F)/P(F)$	conditional probability: probability of E occurring, provided F occurs	
Var x; $\sigma(x)$	variance, root-mean-square deviation of x	
$x_{1/2}$	median	
$x_{1/4}$	first quartile	

$x_{3/4}$		third quartile
$x_{0.01}$		percentile
$\bar{x}$; $<x>$; $E(x)$		expected value of x
α_r		rth moment of density function
$\psi_x(q)$		characteristic function of a distribution
$M_x(s)$		moment-generating function of a distribution
λ_{12}		covariance of two random variables
ρ_{12}		correlation coefficient of two random variables
ξ		expected value and variance of Poisson distribution
R		Spearman's rank correlation
W		coefficient of concordance
R		multiple correlation coefficient
b		regression coefficient
r_t		tetrachoric correlation coefficient
η		correlation ratio

ABBREVIATIONS FOR SCIENTIFIC AND LEARNED SOCIETIES

LEARNED SOCIETIES and professional organizations devoted to scientific and technological advancement are great in number and are highly varied in their activities, often expanding their interest to adjacent areas of study in many directions. This has generated a profusion of technical documents which have brought to the attention of the scientist the imprint of many an organization heretofore virtually unknown out of its own area of specialization. The increasing intercommunication among such organizations has necessitated a comprehensive and authentic reference list whereby the acronyms for these organizations may be readily identified.

The abbreviations given below are those formally approved by their respective societies where such approval has been made; where not, the style most generally used for reference to that society has been followed. Certain societies are organized as either affiliates or subgroups of others; where this is the case, the parent organization is indicated by abbreviation in parenthesis following the name in the entry. The abbreviations used for parent organizations are also to be found in this section. Where a new organization has been formed by the merger of two previously existing groups, those no longer in existence are listed with a parenthetical note indicating the new name.

International organizations have been included whose technical publications are germane to the study of physics or mathematics, or where reference made to them in other literature is frequent. Included also are relevant British and Canadian societies; entries of British societies having names which do not indicate their nationality include the note "[UK]."

ABBREVIATIONS FOR SCIENTIFIC AND LEARNED SOCIETIES
Alphabetically by Abbreviation

AAAS	American Association for the Advancement of Science
AAPT	American Association of Physics Teachers
AAS	American Astronomical Society
ACA	American Crystallographic Association
ACS	American Chemical Society
AFASE	Association for Applied Solar Energy
AGI	American Geological Institute
AGU	American Geophysical Union
AIAA	American Institute of Aeronautics and Astronautics
AIBS	American Institute of Biological Sciences
AIEE	American Institute of Electrical Engineers (merged with IRE to form IEEE)
AIF	Atomic Industrial Forum
AIP	American Institute of Physics
AMS	American Mathematical Society
AMS	American Meteorological Society
ANS	American Nuclear Society
APS	American Physical Society
ARS	American Rocket Society (merged with IAS to form AIAA)
ASA	Acoustical Society of America
ASA	American Standards Association
ASA	American Statistical Association
ASM	American Society for Metals
ASME	American Society of Mechanical Engineers
ASTM	American Society for Testing and Materials
BAAS	British Association for the Advancement of Science
BAS	British Astronomical Society
BIS	British Interplanetary Society
BPS	Biophysical Society
BS	Biometric Society
B.S.I.	British Standards Institution
CCO	Canadian Committee on Oceanography
CFES	Canadian Federation of Engineers and Scientists
CIC	Chemical Institute of Canada
CIE	International Commission on Illumination
COSPAR	Committee on Space Research (ICSU)

DPMA

CS	Chemical Society [UK]
CSA	Canadian Standards Association
ECS	Electrochemical Society
FAGS	Federation of Astronomical and Geophysical Services (ICSU)
FORATOM	European Atomic Forum
GEERS	European Study Group for Space Research
HPS	Health Physics Society
IAB	International Council of Scientific Unions Abstracting Board (ICSU)
IAG	International Association of Geodesy
IAGA	International Association of Geomagnetism and Aeronomy
IAMAP	International Association of Meteorology and Atmospheric Physics
IAPO	International Association of Physical Oceanography (IUGG)
IAS	Institute of the Aerospace Sciences (merged with ARS to form AIAA)
IASPEI	International Association of Seismology and Physics of the Earth's Interior
IAU	International Astronomical Union
ICO	International Commission for Optics
ICRP	International Commission on Radiological Protection
ICRU	International Commission on Radiological Units and Measurements
ICSU	International Council of Scientific Unions
IEC	International Electrotechnical Commission
I.E.E.	Institution of Electrical Engineers [UK]
IEEE	Institute of Electrical and Electronics Engineers
IGC	International Geophysical Committee (ICSU)
I.Mech.E.	Institution of Mechanical Engineers [UK]
IMS	Industrial Mathematics Society
Inst. P.	Institute of Physics [UK]
IOPAB	International Organization for Pure and Applied Biophysics
IRE	Institute of Radio Engineers (merged with AIEE to form IEEE)
ISO	International Organization for Standardization
IUBS	International Union of Biological Sciences
IUCr	International Union of Crystallography

IUGG	International Union of Geodesy and Geophysics
IUGS	International Union of Geological Sciences
IUPAC	International Union of Pure and Applied Chemistry
IUPAP	International Union of Pure and Applied Physics
IUTAM	International Union of Theoretical and Applied Mechanics
MA	Mathematical Association [UK]
MAA	Mathematical Association of America
MURA	Midwestern Universities Research Association
OSA	Optical Society of America
RESA	Scientific Research Society of America
R.Met.S.	Royal Meteorological Society [UK]
RRS	Radiation Research Society
SAS	Society for Applied Spectroscopy
SIAM	Society for Industrial and Applied Mathematics
SSA	Seismological Society of America
SUN Commission	Commission on Symbols, Units and Nomenclature (IUPAP)

ABBREVIATIONS FOR SCIENTIFIC AND LEARNED SOCIETIES
Alphabetically by Name

Acoustical Society of America	ASA
American Association for the Advancement of Science	AAAS
American Association of Physics Teachers	AAPT
American Astronomical Society	AAS
American Chemical Society	ACS
American Crystallographic Association	ACA
American Geological Institute	AGI
American Geophysical Union	AGU
American Institute of Aeronautics and Astronautics	AIAA
American Institute of Biological Sciences	AIBS
American Institute of Electrical Engineers (merged with IRE to form IEEE)	AIEE
American Institute of Physics	AIP
American Mathematical Society	AMS
American Meteorological Society	AMS
American Nuclear Society	ANS
American Physical Society	APS
American Rocket Society (merged with IAS to form AIAA)	ARS
American Society for Metals	ASM
American Society for Testing and Materials	ASTM
American Society of Mechanical Engineers	ASME
American Standards Association	ASA
American Statistical Association	ASA
Association for Applied Solar Energy	AFASE
Atomic Industrial Forum	AIF
Biometric Society	BS
Biophysical Society	BPS
British Association for the Advancement of Science	BAAS
British Astronomical Society	BAS
British Interplanetary Society	BIS
British Standards Institution	B.S.I.
Canadian Committee on Oceanography	CCO
Canadian Federation of Engineers and Scientists	CFES
Canadian Standards Association	CSA
Chemical Institute of Canada	CIC
Chemical Society [UK]	CS

DPMA

Commission on Symbols, Units and Nomenclature (IUPAP)	SUN Commission
Committee on Space Research (ICSU)	COSPAR
Electrochemical Society	ECS
European Atomic Forum	FORATOM
European Study Group for Space Research	GEERS
Federation of Astronomical and Geophysical Services (ICSU)	FAGS
Health Physics Society	HPS
Industrial Mathematics Society	IMS
Institute of Electrical and Electronics Engineers	IEEE
Institute of Physics [UK]	Inst. P.
Institute of Radio Engineers (merged with AIEE to form IEEE)	IRE
Institute of the Aerospace Sciences (merged with ARS to form AIAA)	IAS
Institution of Electrical Engineers [UK]	I.E.E.
Institution of Mechanical Engineers [UK]	I.Mech.E.
International Association of Geodesy	IAG
International Association of Geomagnetism and Aeronomy	IAGA
International Association of Meteorology and Atmospheric Physics	IAMAP
International Association of Physical Oceanography (IUGG)	IAPO
International Association of Seismology and Physics of the Earth's Interior	IASPEI
International Astronomical Union	IAU
International Commission for Optics	ICO
International Commission on Illumination	CIE
International Commission on Radiological Protection	ICRP
International Commission on Radiological Units and Measurements	ICRU
International Council of Scientific Unions	ICSU
International Council of Scientific Unions Abstracting Board (ICSU)	IAB
International Electrotechnical Commission	IEC
International Geophysical Committee (ICSU)	IGC
International Organization for Pure and Applied Biophysics	IOPAB
International Organization for Standardization	ISO

DPMA

International Union of Biological Sciences	IUBS
International Union of Crystallography	IUCr
International Union of Geodesy and Geophysics	IUGG
International Union of Geological Sciences	IUGS
International Union of Pure and Applied Chemistry	IUPAC
International Union of Pure and Applied Physics	IUPAP
International Union of Theoretical and Applied Mechanics	IUTAM
Mathematical Association [UK]	MA
Mathematical Association of America	MAA
Midwestern Universities Research Association	MURA
Optical Society of America	OSA
Radiation Research Society	RRS
Royal Meteorological Society [UK]	R.Met.S.
Scientific Research Society of America	RESA
Seismological Society of America	SSA
Society for Applied Spectroscopy	SAS
Society for Industrial and Applied Mathematics	SIAM

ABBREVIATIONS FOR GOVERNMENT AND MILITARY DEPARTMENTS, AGENCIES, AND OFFICES

MAN'S SCIENTIFIC knowledge has increased more in the seventy years of the twentieth century than in all the previous centuries of his existence. Science has attacked grander and more complex problems, and research has become an increasingly expensive pursuit. However, the fruits of scientific investigation are recognized as a national resource of primary importance, and consequently a large share of the funds supporting scientific research has been made available by the government. Research centers and laboratories exist within a large number of the various branches of the United States Government, and the reason is thus obvious for the constant reference to government and military agencies in scientific publications.

The scientist needs an up-to-date list of the various government and military branches that are repeatedly identified by abbreviation in periodicals, research reports, technical memoranda, and other publications. This section is intended to satisfy that need.

Each subordinate agency is keyed to its parent organization by a parenthetical abbreviation following the definition. For example, the definition of "ARPA" is "Advanced Research Projects Agency (DOD)," where "DOD" is the parent organization, "Department of Defense."

Relevant international, civilian, military and nonprofit research organizations are included.

ABBREVIATIONS FOR GOVERNMENT AND MILITARY DEPARTMENTS, AGENCIES, AND OFFICES
Alphabetically by Abbreviation

ABCC	Atomic Bomb Casualty Commission (AEC)
ACRP	Advisory Committee on Reactor Physics (AEC)
ACRS	Advisory Committee on Reactor Safeguards (AEC)
ACTI	Advisory Committee on Technical Information (AEC)
AEC	Atomic Energy Commission
AFCRC	Air Force Cambridge Research Center
AFCRL	Air Force Cambridge Research Laboratories
AFOAR	Air Force Office of Aerospace Research
AFOSR	Air Force Office of Scientific Research
AMD	Applied Mathematics Division (NBS)
ANL	Argonne National Laboratory (AEC)
ARC	Ames Research Center (NASA)
ARF	Armour Research Foundation (now IITRI)
ARL	Aeronautical Research Laboratory (OAR)
ARPA	Advanced Research Projects Agency (DOD)
ASESA	Armed Services Electro-Standards Agency
ASTIA	Armed Services Technical Information Agency (now DDC)
BAPL	Bettis Atomic Power Laboratory (AEC)
BMI	Battelle Memorial Institute
BNL	Brookhaven National Laboratory (AEC)
CAFEE	Critical Assembly Fuel Element Exchange (AEC)
CANEL	Connecticut Advanced Nuclear Engineering Laboratory (AEC)
CERN	European Organization for Nuclear Research
C&GS	Coast and Geodetic Survey (USCOMM)
CNL	Clinton National Laboratory (AEC)
CNO	Chief of Naval Operations
COESA	Committee on Extension to the Standard Atmosphere (USAF-NASA-USWB)
COPERS	European Preparatory Commission for Space Research (now ESRO)
CRPL	Central Radio Propagation Laboratory (NBS)
DA	Department of the Army
DAF	Department of the Air Force
DDC	Defense Documentation Center for Scientific and Technical Information (DOD)

DPMA

DID	....	Division of Isotopes Development (AEC) ....
DN	....	Department of the Navy ....
DOD	....	Department of Defense ....
DSA	....	Defense Supply Agency (DOD) ....
DTI	....	Division of Technical Information (AEC) ....
EOP	....	Executive Office of the President of the United States ....
ESDAC	....	European Space Data Center (ESRO) ....
ESRIN	....	European Space Research Institute (ESRO) ....
ESRO	....	European Space Research Organization ....
EURATOM	.....	European Atomic Energy Community ....
Eurisotope	......	Radioisotope Information Bureau (EURATOM) ....
FACSI	....	Federal Advisory Committee for Scientific Information (NSF) ....
FCST	....	Federal Council for Science and Technology .....
FI	....	Franklin Institute ....
FRC	....	Federal Radiation Council ....
GPO	....	Government Printing Office ....
GSFC	....	Goddard Space Flight Center (NASA) ....
HASL	....	Health and Safety Laboratory (AEC) ....
HEW	....	Department of Health, Education and Welfare ....
HIRDL	....	High Intensity Radiation Development Laboratory (BNL) ....
HRLEL	....	High Radiation Level Examination Laboratory (ORNL) ....
IAEA	....	International Atomic Energy Agency (UN) ....
IANEC	....	Inter-American Nuclear Energy Commission ......
IBWM	....	International Bureau of Weights and Measures ....
ICO	....	Interagency Committee on Oceanography (FCST) ..
ICWM	....	International Committee on Weights and Measures ....
IITRI	....	Illinois Institute of Technology Research Institute ..
IOC	....	Intergovernmental Oceanographic Commission (UNESCO) ....
ITU	....	International Telecommunication Union (UN) ....
JCAE	....	Joint Committee on Atomic Energy (Congress) ....
JILA	....	Joint Institute for Laboratory Astrophysics (NBS) ..
JPL	....	Jet Propulsion Laboratory (NASA) ....
KAPL	....	Knolls Atomic Power Laboratory (AEC) ....
LaRC	....	Langley Research Center (NASA) ....
LASL	....	Los Alamos Scientific Laboratory (AEC) ....
LC	....	Library of Congress ....

LRC	Lewis Research Center (NASA)
LRL	Lawrence Radiation Laboratory (AEC)
MADAEC	Military Application Division of the Atomic Energy Commission
MITRE	Massachusetts Institute of Technology Research and Engineering Corporation
MSC	Manned Spacecraft Center (NASA)
MSFC	Marshall Space Flight Center (NASA)
NAS	National Academy of Sciences
NASA	National Aeronautics and Space Administration
NASCO	National Academy of Sciences Committee on Oceanography
NBS	National Bureau of Standards (USCOMM)
NCAR	National Center for Atmospheric Research (NSF)
NEL	Naval Electronics Laboratory (USN)
NMC	National Meteorological Center (USWB)
NOL	Naval Ordnance Laboratory (USN)
NRC	National Research Council (NAS)
NRL	Naval Research Laboratory (USN)
NRTS	National Reactor Testing Station (AEC)
NSF	National Science Foundation
NSIC	Nuclear Safety Information Center (AEC)
NWSC	National Weather Satellite Center (USWB)
OAR	Office of Aerospace Research (USAF)
OART	Office of Advanced Research and Technology (NASA)
OE	Office of Education (HEW)
ONR	Office of Naval Research (CNO)
ORINS	Oak Ridge Institute of Nuclear Studies (AEC)
ORNL	Oak Ridge National Laboratory (AEC)
ORSORT	Oak Ridge School of Reactor Technology (AEC)
OSIS	Office of Science Information Service (NSF)
OSR	Office of Scientific Research (USAF)
OSRD	Office of Scientific Research and Development (DOD)
OSS	Office of Space Sciences (NASA)
OST	Office of Science and Technology (EOP)
OTS	Office of Technical Services (USCOMM)
REIC	Radiation Effects Information Center (BMI)
SAB	Scientific Advisory Board (USAF)
SAO	Smithsonian Astrophysical Observatory
SEIC	Solar Energy Information Center (DOD)

SigC	Signal Corps (USA)
SIO	Scripps Institution of Oceanography
SRDC	Standard Reference Data Center (NBS)
SRI	Stanford Research Institute
STL	Space Technology Laboratories (NASA)
UCLRL	University of California Lawrence Radiation Laboratory (AEC)
UN	United Nations
UNESCO	United Nations Educational, Scientific and Cultural Organization
UNSCEAR	United Nations Scientific Committee on the Effects of Atomic Radiation
USA	United States Army
USAEC	United States Atomic Energy Commission
USAF	United States Air Force
USC&GS	United States Coast and Geodetic Survey (USCOMM)
USCOMM	United States Department of Commerce
USN	United States Navy
USOE	United States Office of Education (HEW)
USWB	United States Weather Bureau (USCOMM)
WB	Weather Bureau (USCOMM)
WHOI	Woods Hole Oceanographic Institution
WMO	World Meteorological Organization (UN)

ABBREVIATIONS FOR GOVERNMENT AND MILITARY DEPARTMENTS, AGENCIES, AND OFFICES
Alphabetically by Name

Advanced Research Projects Agency (DOD) ...... **ARPA** ....
Advisory Committee on Reactor Physics (AEC) .... **ACRP** ....
Advisory Committee on Reactor Safeguards (AEC) . **ACRS** ....
Advisory Committee on Technical Information (AEC) .... **ACTI** ....
Aeronautical Research Laboratory (OAR) .... **ARL** ....
Air Force Cambridge Research Center .... **AFCRC** ....
Air Force Cambridge Research Laboratories ...... **AFCRL** ....
Air Force Office of Aerospace Research .... **AFOAR** ....
Air Force Office of Scientific Research .... **AFOSR** ....
Ames Research Center (NASA) .... **ARC** ....
Applied Mathematics Division (NBS) .... **AMD** ....
Argonne National Laboratory (AEC) .... **ANL** ....
Armed Services Electro-Standards Agency .... **ASESA** ....
Armed Services Technical Information Agency (now DDC) .... **ASTIA** ....
Armour Research Foundation (now IITRI) .... **ARF** ....
Atomic Bomb Casualty Commission (AEC) .... **ABCC** ....
Atomic Energy Commission .... **AEC** ....
Battelle Memorial Institute .... **BMI** ....
Bettis Atomic Power Laboratory (AEC) .... **BAPL** ....
Brookhaven National Laboratory (AEC) .... **BNL** ....
Central Radio Propagation Laboratory (NBS) ...... **CRPL** ....
Chief of Naval Operations .... **CNO** ....
Clinton National Laboratory (AEC) .... **CNL** ....
Coast and Geodetic Survey (USCOMM) .... **C&GS** ....
Committee on Extension to the Standard Atmosphere (USAF-NASA-USWB) .... **COESA** ....
Connecticut Advanced Nuclear Engineering Laboratory (AEC) .... **CANEL** ....
Critical Assembly Fuel Element Exchange (AEC) ... **CAFEE** ....
Defense Documentation Center for Scientific and Technical Information (DOD) .... **DDC** ....
Defense Supply Agency (DOD) .... **DSA** ....
Department of Defense .... **DOD** ....
Department of Health, Education and Welfare .... **HEW** ....
Department of the Air Force .... **DAF** ....
Department of the Army .... **DA** ....

DPMA

Department of the Navy	DN
Division of Isotopes Development (AEC)	DID
Division of Technical Information (AEC)	DTI
European Atomic Energy Community	EURATOM
European Organization for Nuclear Research	CERN
European Preparatory Commission for Space Research (now ESRO)	COPERS
European Space Data Center (ESRO)	ESDAC
European Space Research Institute (ESRO)	ESRIN
European Space Research Organization	ESRO
Executive Office of the President of the United States	EOP
Federal Advisory Committee for Scientific Information (NSF)	FACSI
Federal Council for Science and Technology	FCST
Federal Radiation Council	FRC
Franklin Institute	FI
Goddard Space Flight Center (NASA)	GSFC
Government Printing Office	GPO
Health and Safety Laboratory (AEC)	HASL
High Intensity Radiation Development Laboratory (BNL)	HIRDL
High Radiation Level Examination Laboratory (ORNL)	HRLEL
Illinois Institute of Technology Research Institute	IITRI
Interagency Committee on Oceanography (FCST)	ICO
Inter-American Nuclear Energy Commission	IANEC
Intergovernmental Oceanographic Commission (UNESCO)	IOC
International Atomic Energy Agency (UN)	IAEA
International Bureau of Weights and Measures	IBWM
International Committee on Weights and Measures	ICWM
International Telecommunication Union (UN)	ITU
Jet Propulsion Laboratory (NASA)	JPL
Joint Committee on Atomic Energy (Congress)	JCAE
Joint Institute for Laboratory Astrophysics (NBS)	JILA
Knolls Atomic Power Laboratory (AEC)	KAPL
Langley Research Center (NASA)	LaRC
Lawrence Radiation Laboratory (AEC)	LRL
Lewis Research Center (NASA)	LRC
Library of Congress	LC

Los Alamos Scientific Laboratory (AEC) .... **LASL** ....
Manned Spacecraft Center (NASA) .... **MSC** ....
Marshall Space Flight Center (NASA) .... **MSFC** ....
Massachusetts Institute of Technology Research
   and Engineering Corporation .... **MITRE** ....
Military Application Division of the Atomic
   Energy Commission .... **MADAEC** ......
National Academy of Sciences .... **NAS** ....
National Academy of Sciences Committee on
   Oceanography .... **NASCO** ....
National Aeronautics and Space Administration ... **NASA** ....
National Bureau of Standards (USCOMM) .... **NBS** ....
National Center for Atmospheric Research (NSF) .. **NCAR** ....
National Meteorological Center (USWB) .... **NMC** ....
National Reactor Testing Station (AEC) .... **NRTS** ....
National Research Council (NAS) .... **NRC** ....
National Science Foundation .... **NSF** ....
National Weather Satellite Center (USWB) .... **NWSC** ....
Naval Electronics Laboratory (USN) .... **NEL** ....
Naval Ordnance Laboratory (USN) .... **NOL** ....
Naval Research Laboratory (USN) .... **NRL** ....
Nuclear Safety Information Center (AEC) .... **NSIC** ....
Oak Ridge Institute of Nuclear Studies (AEC) ..... **ORINS** ....
Oak Ridge National Laboratory (AEC) .... **ORNL** ....
Oak Ridge School of Reactor Technology (AEC) .. **ORSORT** ....
Office of Advanced Research and Technology
   (NASA) .... **OART** ....
Office of Aerospace Research (USAF) .... **OAR** ....
Office of Education (HEW) .... **OE** ....
Office of Naval Research (CNO) .... **ONR** ....
Office of Science and Technology (EOP) .... **OST** ....
Office of Science Information Service (NSF) ...... **OSIS** ....
Office of Scientific Research (USAF) .... **OSR** ....
Office of Scientific Research and Development
   (DOD) .... **OSRD** ....
Office of Space Sciences (NASA) .... **OSS** ....
Office of Technical Services (USCOMM) .... **OTS** ....
Radiation Effects Information Center (BMI) .... **REIC** ....
Radioisotope Information Bureau (EURATOM) .... **Eurisotope** ......
Scientific Advisory Board (USAF) .... **SAB** ....
Scripps Institution of Oceanography .... **SIO** ....
Signal Corps (USA) .... **SigC** ....

Smithsonian Astrophysical Observatory	SAO
Solar Energy Information Center (DOD)	SEIC
Space Technology Laboratories (NASA)	STL
Standard Reference Data Center (NBS)	SRDC
Stanford Research Institute	SRI
United Nations	UN
United Nations Educational, Scientific and Cultural Organization	UNESCO
United Nations Scientific Committee on the Effects of Atomic Radiation	UNSCEAR
United States Air Force	USAF
United States Army	USA
United States Atomic Energy Commission	USAEC
United States Coast and Geodetic Survey (USCOMM)	USC&GS
United States Department of Commerce	USCOMM
United States Navy	USN
United States Office of Education (HEW)	USOE
United States Weather Bureau (USCOMM)	USWB
University of California Lawrence Radiation Laboratory (AEC)	UCLRL
Weather Bureau (USCOMM)	WB
Woods Hole Oceanographic Institution	WHOI
World Meteorological Organization (UN)	WMO

TABLE OF CHEMICAL ELEMENTS

Table reflects caution on accuracy
Based on the assigned relative atomic mass of $^{12}C = 12$

The following values apply to elements as they exist in materials of terrestrial origin and to certain artificial elements. When used with the footnotes, they are reliable to ± 1 in the last digit, or ± 3 if that digit is in small type

Symbol	Element	Atomic number	Atomic weight
Ac	Actinium	89	
Al	Aluminum	13	26.9815[a]
Am	Americium	95	
Sb	Antimony	51	121.75
Ar	Argon	18	39.948[b,c,d,g]
As	Arsenic	33	74.9216[a]
At	Astatine	85	
Ba	Barium	56	137.34
Bk	Berkelium	97	
Be	Beryllium	4	9.01218[a]
Bi	Bismuth	83	208.9806[a]
B	Boron	5	10.81[c,d,e]
Br	Bromine	35	79.904[c]
Cd	Cadmium	48	112.40
Ca	Calcium	20	40.08
Cf	Californium	98	
C	Carbon	6	12.011[b,d]
Ce	Cerium	58	140.12
Cs	Cesium	55	132.9055[a]
Cl	Chlorine	17	35.453[c]
Cr	Chromium	24	51.996[c]
Co	Cobalt	27	58.9332[a]
Cu	Copper	29	63.546[c,d]
Cm	Curium	96	
Dy	Dysprosium	66	162.50

[a] Mononuclidic element.
[b] Element with one predominant isotope (about 99 to 100% abundance).
[c] Element for which the atomic weight is based on calibrated measurements.
[d] Element for which variation in isotopic abundance in terrestrial samples limits the precision of the atomic weight given.
[e] Element for which users are cautioned against the possibility of large variations in atomic weight due to inadvertent or undisclosed artificial isotopic separation in commercially available materials.
[f] Most commonly available long-lived isotope; see a Table of Selected Radioactive Isotopes.
[g] In some geological specimens this element has a highly anomalous isotopic composition, corresponding to an atomic weight significantly different from that given.

DPMA

Symbol	Element	Atomic number	Atomic weight
Es	Einsteinium	99	
Er	Erbium	68	167.26
Eu	Europium	63	151.96
Fm	Fermium	100	
F	Fluorine	9	18.9984[a]
Fr	Francium	87	
Gd	Gadolinium	64	157.25
Ga	Gallium	31	69.72
Ge	Germanium	32	72.59
Au	Gold	79	196.9665[a]
Hf	Hafnium	72	178.49
He	Helium	2	4.00260[b,e]
Ho	Holmium	67	164.9303[a]
H	Hydrogen	1	1.0080[b,d]
In	Indium	49	114.82
I	Iodine	53	126.9045[a]
Ir	Iridium	77	192.22
Fe	Iron	26	55.847
Kr	Krypton	36	83.80
La	Lanthanum	57	138.9055[b]
Lr	Lawrencium	103	
Pb	Lead	82	207.2[d,g]
Li	Lithium	3	6.941[c,d,e]
Lu	Lutetium	71	174.97
Mg	Magnesium	12	24.305[c]
Mn	Manganese	25	54.9380[a]
Md	Mendelevium	101	
Hg	Mercury	80	200.59
Mo	Molybdenum	42	95.94
Nd	Neodymium	60	144.24
Ne	Neon	10	20.179[e]
Np	Neptunium	93	237.0482[b]
Ni	Nickel	28	58.71
Nb	Niobium	41	92.9064[a]
N	Nitrogen	7	14.0067[b,c]
No	Nobelium	102	
Os	Osmium	76	190.2
O	Oxygen	8	15.9994[b,c,d]
Pd	Palladium	46	106.4
P	Phosphorus	15	30.9738[a]

Symbol	Element	Atomic number	Atomic weight
Pt	Platinum	78	195.0$_9$
Pu	Plutonium	94	
Po	Polonium	84	
K	Potassium	19	39.10$_2$
Pr	Praseodymium	59	140.0977[a]
Pm	Promethium	61	
Pa	Protactinium	91	231.0359[a]
Ra	Radium	88	226.0254[a,f,g]
Rn	Radon	86	
Re	Rhenium	75	186.2
Rh	Rhodium	45	102.9055[a]
Rb	Rubidium	37	85.4678[c]
Ru	Ruthenium	44	101.0$_7$
Sm	Samarium	62	150.4
Sc	Scandium	21	44.9559[a]
Se	Selenium	34	78.9$_6$
Si	Silicon	14	28.086[d]
Ag	Silver	47	107.868[c]
Na	Sodium	11	22.9898[a]
Sr	Strontium	38	87.62[g]
S	Sulfur	16	32.06[d]
Ta	Tantalum	73	180.9479[b]
Tc	Technetium	43	98.9062[f]
Te	Tellurium	52	127.6$_0$
Tb	Terbium	65	158.9254[a]
Tl	Thallium	81	204.3$_7$
Th	Thorium	90	232.0381[a]
Tm	Thulium	69	168.9342[a]
Sn	Tin	50	118.6$_9$
Ti	Titanium	22	47.9$_0$
W	Tungsten	74	183.8$_5$
U	Uranium	92	238.029[b,c,e]
V	Vanadium	23	50.9414[b,e]
W	Wolfram	74	183.8$_5$
Xe	Xenon	54	131.30
Yb	Ytterbium	70	173.04
Y	Yttrium	39	88.9059[a]
Zn	Zinc	30	65.3$_7$
Zr	Zirconium	40	91.22

PERIODIC TABLE OF THE ELEMENTS

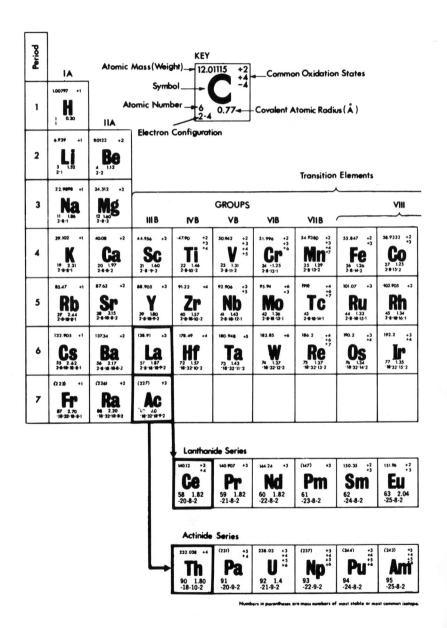

GROUPS

		IB	IIB	IIIA	IVA	VA	VIA	VIIA	INERT GASES 0	ORBIT
								1.00797 −1 $\;$ H $\;$ 1 $\;$ 0.30	4.0026 $\;$ 0 $\;$ He $\;$ 2 $\;$ 0.93	K
				10.811 +3 $\;$ B $\;$ 5 $\;$ 0.88 $\;$ 2-3	12.01115 +2 +4 −4 $\;$ C $\;$ 6 $\;$ 0.77 $\;$ 2-4	14.0067 +1 +2 +3 +4 +5 −1 −2 −3 $\;$ N $\;$ 7 $\;$ 0.70 $\;$ 2-5	15.9994 −2 $\;$ O $\;$ 8 $\;$ 0.66 $\;$ 2-6	18.9984 −1 $\;$ F $\;$ 9 $\;$ 0.64 $\;$ 2-7	20.183 $\;$ 0 $\;$ Ne $\;$ 10 $\;$ 1.12 $\;$ 2-8	K-L
				26.9815 +3 $\;$ Al $\;$ 13 $\;$ 1.17 $\;$ 2-8-3	28.086 +2 +4 −4 $\;$ Si $\;$ 14 $\;$ 1.17 $\;$ 2-8-4	30.9738 +3 +5 −3 $\;$ P $\;$ 15 $\;$ 1.10 $\;$ 2-8-5	32.064 +4 +6 −2 $\;$ S $\;$ 16 $\;$ 1.04 $\;$ 2-8-6	35.453 +1 +5 +7 −1 $\;$ Cl $\;$ 17 $\;$ 0.99 $\;$ 2-8-7	39.948 $\;$ 0 $\;$ Ar $\;$ 18 $\;$ 1.54 $\;$ 2-8-8	K-L-M
58.71 +2 +3 $\;$ Ni $\;$ 28 $\;$ 1.24 $\;$ 2-8-16-2	63.54 +1 +2 $\;$ Cu $\;$ 29 $\;$ 1.28 $\;$ 2-8-18-1	65.37 +2 $\;$ Zn $\;$ 30 $\;$ 1.33 $\;$ 2-8-18-2	69.72 +3 $\;$ Ga $\;$ 31 $\;$ 1.22 $\;$ 2-8-18-3	72.59 +2 +4 $\;$ Ge $\;$ 32 $\;$ 1.22 $\;$ 2-8-18-4	74.9216 +3 +5 −3 $\;$ As $\;$ 33 $\;$ 1.21 $\;$ 2-8-18-5	78.96 +4 +6 −2 $\;$ Se $\;$ 34 $\;$ 1.17 $\;$ 2-8-18-6	79.909 +1 +5 −1 $\;$ Br $\;$ 35 $\;$ 1.14 $\;$ 2-8-18-7	83.80 $\;$ 0 $\;$ Kr $\;$ 36 $\;$ 1.69 $\;$ 2-8-18-8	-L-M-N
106.4 +2 +4 $\;$ Pd $\;$ 46 $\;$ 1.38 $\;$ 2-8-18-18	107.870 +1 $\;$ Ag $\;$ 47 $\;$ 1.44 $\;$ 2-8-18-18-1	112.40 +2 $\;$ Cd $\;$ 48 $\;$ 1.49 $\;$ 2-8-18-18-2	114.82 +3 $\;$ In $\;$ 49 $\;$ 1.62 $\;$ 2-8-18-18-3	118.69 +2 +4 $\;$ Sn $\;$ 50 $\;$ 1.40 $\;$ 2-8-18-18-4	121.75 +3 +5 −3 $\;$ Sb $\;$ 51 $\;$ 1.41 $\;$ 2-8-18-18-5	127.60 +4 +6 −2 $\;$ Te $\;$ 52 $\;$ 1.37 $\;$ 2-8-18-18-6	126.9044 +1 +5 +7 −1 $\;$ I $\;$ 53 $\;$ 1.33 $\;$ 2-8-18-18-7	131.30 $\;$ 0 $\;$ Xe $\;$ 54 $\;$ 1.90 $\;$ 2-8-18-18-8	-M-N-O
195.09 +2 +4 $\;$ Pt $\;$ 78 $\;$ 1.38 $\;$ -18-32-17-1	196.967 +1 +2 $\;$ Au $\;$ 79 $\;$ 1.44 $\;$ -18-32-18-1	200.59 +1 +2 $\;$ Hg $\;$ 80 $\;$ 1.55 $\;$ -18-32-18-2	204.37 +1 +3 $\;$ Tl $\;$ 81 $\;$ 1.71 $\;$ -18-32-18-3	207.19 +2 +4 $\;$ Pb $\;$ 82 $\;$ 1.75 $\;$ -18-32-18-4	208.980 +3 +5 $\;$ Bi $\;$ 83 $\;$ 1.46 $\;$ -18-32-18-5	(210) +2 +4 $\;$ Po $\;$ 84 $\;$ 1.65 $\;$ -18-32-18-6	(210) $\;$ At $\;$ 85 $\;$ 1.40 $\;$ -18-32-18-7	(222) $\;$ 0 $\;$ Rn $\;$ 86 $\;$ 2.2 $\;$ -18-32-18-8	-N-O-P
										O-P-Q

157.25 +3 $\;$ Gd $\;$ 64 $\;$ 1.79 $\;$ -25-9-2	158.924 +3 $\;$ Tb $\;$ 65 $\;$ 1.77 $\;$ -27-8-2	162.50 +3 $\;$ Dy $\;$ 66 $\;$ 1.77 $\;$ -28-8-2	164.930 +3 $\;$ Ho $\;$ 67 $\;$ 1.76 $\;$ -29-8-2	167.26 +3 $\;$ Er $\;$ 68 $\;$ 1.75 $\;$ -30-8-2	168.934 +3 $\;$ Tm $\;$ 69 $\;$ 1.74 $\;$ -31-8-2	173.04 +2 +3 $\;$ Yb $\;$ 70 $\;$ 1.93 $\;$ -32-8-2	174.97 +3 $\;$ Lu $\;$ 71 $\;$ 1.74 $\;$ -32-9-2	N-O-P

(247) +3 $\;$ Cm $\;$ 96 $\;$ -25-9-2	(247) +3 +4 $\;$ Bk $\;$ 97 $\;$ -27-8-2	(251) +3 $\;$ Cf $\;$ 98 $\;$ -28-8-2	(254) $\;$ Es $\;$ 99 $\;$ -29-8-2	(257) $\;$ Fm $\;$ 100 $\;$ -30-8-2	(256) $\;$ Md $\;$ 101 $\;$ -31-8-2	(254) $\;$ No $\;$ 102 $\;$ -32-8-2	(257) $\;$ Lw $\;$ 103 $\;$ -32-9-2	— $\;$ 104 $\;$ -32-10-2	-O-P-Q

SOURCES of STANDARDS and DOCUMENTS

The material reported in this Dictionary has been compiled by reference to the standards, publications, and personnel of the following:

Acoustical Society of America
American Association for the Advancement of Science
American Astronomical Society
American Chemical Society
American Geophysical Union
American Institute of Physics
American Mathematical Society
American Nuclear Society
American Physical Society
American Society of Mechanical Engineers
American Standards Association
American Statistical Association
Ames Research Center, National Aeronautics and Space Administration
Argonne National Laboratory, United States Atomic Energy Commission
Astrophysical Observatory, Smithsonian Institution
Battelle Memorial Institute
British Standards Institution
Brookhaven National Laboratory, United States Atomic Energy Commission
Canadian Standards Association
Defense Documentation Center for Scientific and Technical Information, Department of Defense
Department of Defense
Department of Health, Education and Welfare
Department of the Air Force
Department of the Army
Department of the Navy
European Nuclear Energy Agency
European Organization for Nuclear Research
Federal Council for Science and Technology
Federal Radiation Council
Goddard Space Flight Center, National Aeronautics and Space Administration

Gravity Research Foundation
Illinois Institute of Technology Research Institute
Institute of Electrical and Electronics Engineers
Institute of Physics and the Physical Society [UK]
Institution of Electrical Engineers [UK]
International Atomic Energy Agency
International Bureau of Weights and Measures
International Commission on Radiological Units and Measurements
International Electrotechnical Commission
International Organization for Standardization
International Union of Pure and Applied Chemistry
International Union of Pure and Applied Physics
Knolls Atomic Power Laboratory, United States Atomic Energy Commission
Langley Research Center, National Aeronautics and Space Administration
Lawrence Radiation Laboratory, United States Atomic Energy Commission
Lewis Research Center, National Aeronautics and Space Administration
Library of Congress
Los Alamos Scientific Laboratory, United States Atomic Energy Commission
Manned Spacecraft Center, National Aeronautics and Space Administration
Marshall Space Flight Center, National Aeronautics and Space Administration
Mathematics Research Center, United States Army
National Aeronautics and Space Administration
National Bureau of Standards, Department of Commerce
National Research Council, National Academy of Sciences
National Science Foundation
Oak Ridge Institute of Nuclear Studies, United States Atomic Energy Commission
Office of Aerospace Research, United States Air Force
Office of Education, Department of Health, Education and Welfare
Office of Naval Research, United States Navy
Office of Science Information Service, National Science Foundation
Office of Scientific Research, United States Air Force
Office of Technical Services, Department of Commerce
Scientific Research Society of America
Society for Industrial and Applied Mathematics

DPMA

Standardization Division, Defense Supply Agency
The Royal Society [UK]
United States Army Signal Corps
United States Atomic Energy Commission
United States Committee on Extension to the Standard Atmosphere
United States Government Printing Office

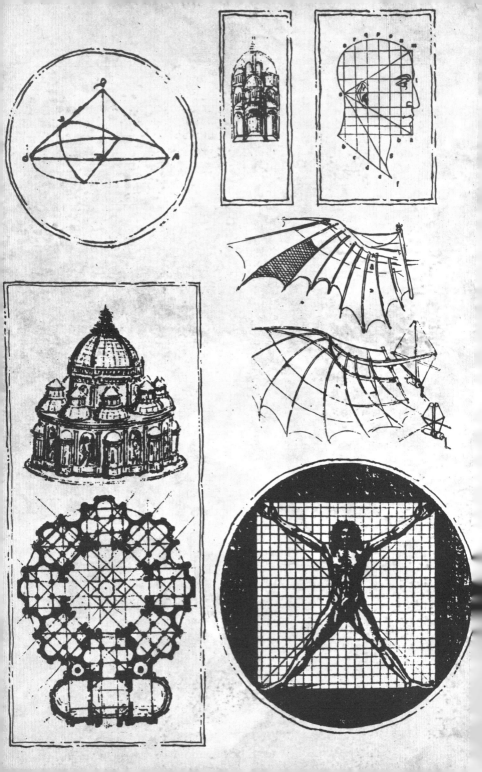